T0252869

Sad by Design

Digital Barricades:
Interventions in Digital Culture and Politics

Series editors:
Professor Jodi Dean, Hobart and William Smith Colleges
Dr Joss Hands, Newcastle University
Professor Tim Jordan, University of Sussex

Also available:

Shooting a Revolution:
Visual Media and Warfare in Syria
Donatella Della Ratta

Cyber-Proletariat:
Global Labour in the Digital Vortex
Nick Dyer-Witheford

The Digital Party:
Political Organisation and Online Democracy
Paolo Gerbaudo

Gadget Consciousness:
Collective Thought, Will and Action in the Age of Social Media
Joss Hands

Information Politics:
Liberation and Exploitation in the Digital Society
Tim Jordan

Unreal Objects:
Digital Materialities, Technoscientific Projects and Political Realities
Kate O'Riordan

Sad by Design

On Platform Nihilism

Geert Lovink

PLUTO  PRESS

First published 2019 by Pluto Press
345 Archway Road, London N6 5AA

www.plutobooks.com

Copyright © Geert Lovink 2019

The right of Geert Lovink to be identified as the author of this work has
been asserted by him in accordance with the Copyright, Designs and
Patents Act 1988.

British Library Cataloguing in Publication Data
A catalogue record for this book is available from the British Library

ISBN   978 0 7453 3935 1   Hardback
ISBN   978 0 7453 3934 4   Paperback
ISBN   978 1 7868 0451 8   PDF eBook
ISBN   978 1 7868 0453 2   Kindle eBook
ISBN   978 1 7868 0452 5   EPUB eBook

This book is printed on paper suitable for recycling and made from fully
managed and sustained forest sources. Logging, pulping and manufacturing
processes are expected to conform to the environmental standards of the
country of origin.

Typeset by Stanford DTP Services, Northampton, England

Printed and bound by CPI Group (UK) Ltd, Croydon, CR0 4YY

Contents

Series Preface

Crisis and conflict open up opportunities for liberation. In the early twenty-first century, these moments are marked by struggles enacted over and across the boundaries of the virtual, the digital, the actual, and the real. Digital cultures and politics connect people even as they simultaneously place them under surveillance and allow their lives to be mined for advertising. This series aims to intervene in such cultural and political conjunctures. It features critical explorations of the new terrains and practices of resistance, producing critical and informed explorations of the possibilities for revolt and liberation.

Emerging research on digital cultures and politics investigates the effects of the widespread digitization of increasing numbers of cultural objects, the new channels of communication swirling around us and the changing means of producing, remixing and distributing digital objects. This research tends to oscillate between agendas of hope, that make remarkable claims for increased participation, and agendas of fear, that assume expanded repression and commodification. To avoid the opposites of hope and fear, the books in this series aggregate around the idea of the barricade. As sources of enclosure as well as defenses for liberated space, barricades are erected where struggles are fierce and the stakes are high. They are necessarily partisan divides, different politicizations and deployments of a common surface. In this sense, new media objects, their networked circuits and settings, as well as their material, informational, and biological carriers all act as digital barricades.

Jodi Dean, Joss Hands and Tim Jordan

Acknowledgements

Three years have passed since I handed in the manuscript of *Social Media Abyss*. Looking back, this period has been defined by Brexit, Trump, Cambridge Analytica and related Facebook scandals. Absent were the equivalents of Wikileaks, Anonymous and Snowden (perhaps with the exception of Cambridge Analytica whistle-blower Chris Wylie). There was a lot of ideological regression and "multi-polar" stagnation, accelerated by a booming stock market, real estate excesses, an ongoing war in Syria and real existing climate change. While the "social media debate" hit the mainstream, resistance and alternatives were all but absent—with the exception of the alt-right.

As the public discourse around "behavior modification" and "pervasive surveillance" intensified, I shifted away from the preferred panorama format used in my previous books. Dutch cultural funding for new projects at our Institute of Network Cultures had dried up (mainly due to new Dutch apartheid walls erected between education/research and the cultural sector). This forced us to go into survival mode and work for third parties, which is the main reason I'm not reporting about the activities of our networks here (even though MoneyLab stills runs and various digital publishing experiments continue, despite the closure of the PublishingLab in 2018).

Thanks to David Castle at Pluto Press and the editors of the Digital Barricades series for taking this book project on board and for their valuable comments.

Working together with copy editor Luke Munn (Auckland) was an amazingly smooth and most pleasurable experience.

My Berlin dialogues over the years with Pit Schultz, Michael Seemann, Volker Grassmuck, Andreas Kallfelz, Stefan Heidenreich and Alexander Karschnia have proven vital for my intellectual wellbeing. Gratitude goes to Steven Shapiro, Ana Peraica, Marc Tuters, Mieke Gerritzen, Franco Berardi, Letizia Chiappini, Daniel de Zeeuw, Ellen Rutten, Alex Foti, Isabel de Maurissens, Morris Kolman, Marie Lechner, Florian Cramer, Katharina Teichgräber, Michael Dieter, Tripta Chandola, Tatjana Seitz and Donatella della Ratta for their inspiring comments on different chapters.

At the Institute of Network Cultures I would like to thank in particular Miriam Rasch, Inte Gloerich and Patricia de Vries for their detailed responses and support. Thanks also to Silvio Lorusso, Leonieke van Dipten, Margreet Riphagen, and to Barbara Dubbeldam and Kelly Mostert, who recently joined us. The entire team has been pivotal in running such an inspiring, critical research center.

Special thanks go to Ned Rossiter for his friendship and dedicated feedback over the years—and to Linda and Riddim DJ "System of Chaos" Kazimir for their amazing support, to whom I dedicate this book.

Amsterdam, October 2018
Geert Lovink

Introduction: Society of the Social

"There are 87,146 thought leaders on LinkedIn."—"Real painters do not paint things as they are…They paint them as they themselves feel them to be." Vincent van Gogh on fake art—Unload that Truck of Dislikes! (alt-left slogan)—"Web: We noticed you're using an Ad blocker. Me: I noticed you're using 32 tracking services." Matt Weagle—"New Security Comes With New Vulnerability." Lulzsec—"Truth is for suckers, Johnny Boy." Being John Malkovich—Our focus is the cosmo-technics challenge that brings us in direct contact with our slaves (tribute to Yuk Hui)—"I always knew I was a good writer but I thought I'd do poetry or fiction, not the emails I ended up doing." OH—"Das Handy und die Zuhandenheit des Virtuellen" (German essay)—"One of my favorite self-harm techniques is googling airfares to Bali." Addie Wagenknecht—"It's not size, it's scale that counts." Barnett Newman—"Warning: People might not like you after this."—"Smart is the new smoking." Johanna Sjerpstra—"Please like our DNS poisoning attack here"—"I HAVE THE HOUSE TO MYSELF TONIGHT! *stares at phone*"—"The Internet is like the Wild West. We thought we were the cowboys, but it turns out we're the buffalos." AnthroPunk.

Welcome to the New Normal. Social media is reformatting our interior lives. As platform and individual become inseparable, social networking becomes identical with the "social" itself. No longer curious about what "the next web" will bring, we chat about the information we're allowed to graze on during meager days. Forward-looking confidence has been shattered—the seasonality of hype reduced to a flatline future. Instead, a new realism has set in, as Evgeny Morozov tweeted: "1990s tech utopianism posited that networks weaken or replace hierarchies. In reality, networks amplify hierarchies and make them less visible."[1] How can one write a proper phenomenology of asynchronous connections and their cultural effects, formulate a ruthless critique of everything hardwired into the social body of the network, while not looking at what's going on inside? Rather than a stance of superiority, a judgment from on high, could we take an amoral approach toward today's intense social media usage, delving into the shallow time of lost souls like us? Let's embark on a journey into this third space called the techno-social.

Our beloved internet may be portrayed as an "inverse hydra with a hundred assholes"[2] but we love it anyway: it's our brain-junk. While social media controversies have hit mainstream media, the fallout has been zero. We barely register the online frenzy that surrounds us; we can't even pretend to care about the cynical advertisement logic.[3] Social media scandals appear to us, as Franz Kafka once wrote, "like a path in autumn: no sooner is it cleared than it is once again littered with leaves." From behavioral manipulation to fake news, all we read about is the bankrupt credibility of Silicon Valley.

However, very few have suffered any serious consequences. Evidence is apparently not enough. Muck gets raked, data gets leaked, and whistles get blown—yet nothing changes. None of the outstanding issues get resolved. There's no "internexit" referendum ahead. No matter how many hacks and privacy violations occur, no matter how many awareness campaigns and public debates are organized, overwhelming indifference prevails. Witness the rapid return to normal following the March 2018 Cambridge Analytica scandal. The centralization of infrastructure and services that provide us with so much comfort is seen as inevitable, ineluctable even.[4] Why aren't there already viable alternatives to the main platforms? Someday we'll understand the Digital Thermidor—but that "someday" never comes.

What's the fate of critique without consequences? As Franco Berardi explained to me when I visited him in Bologna to discuss this book project, it is truth that makes us sad. We lack role models and heroes. Instead we have paranoid truth-seekers. As our responses to the alt-right and systemic violence are so predictable and powerless, Franco suggested to me that we should stop speaking. No reply. Refuse to become news. Do not feed the trolls. The techno-sadness, as explained in this book, has no end, it's bottomless.

How do we reverse the acceleration of alienation, a movement that inevitably ends up in trauma? Instead of pathetic, empty gestures, we should exercise a new tactic of silence, directing the freed energy and resources toward creating temporary spaces of reflection.

In his 2018 book *Anti-Social Media: How Facebook Disconnects Us and Undermines Democracy*, Siva Vaidhyanathan struggles with the growing gap between good intentions and the ugly reality: "The painful paradox of Facebook is that the company's sincere devotion to making the world better invited nefarious parties to hijack it to spread hatred and confusion. Zuckerberg's firm belief in his own expertise, authority, and ethical core blinded him and his company to the damage it was facilitating and causing. If Facebook had been less obsessed with making the world better, it might have avoided contributing to forces that have made the world worse."[5] See here the real existing stagnation, now that the world is digitized. As Gramsci said, "the old is dying and the new cannot be born; in this interregnum a great variety of morbid symptoms appear."

On paper our global challenges look enormous; on screen they fail to be translated into our everyday life. Instead of facing the titanic forces right in the eye, we're numbed, bittersweet, absent-minded, quirky, and sometimes straight-out depressed. Should we read intense social media usage as a coping mechanism? Ours is a profoundly non-heroic, non-mythological, straight-out flat era. After all, myths are stories that need time to develop a broad audience, to ramp up their tension, and to play out their drama. No, our time is marked by the micro concerns of the fragile self. Everyone has his or her reason to shut down and shield off. While corporations can grow overnight to become behemoth structures, outlandish in their infrastructure, our understanding of the world lags behind or even shrinks.

Limited understanding limits our ability to frame the problem. We are not sick.[6] Alarmism has worn itself out. If we want to smash platform capitalism, a political economy analysis will not be sufficient. How might we construct a collective identity, a self-hermeneutics we can live with? Indeed, what would a self-image even be that went beyond machine-readable interpretations? The selfie as mask? "I love that one with you, wearing sunglasses, when you proudly smile." Unable to pin down a problem or articulate a response, the irresistible lure of swiping, updates and "Likes" seems stronger than ever. Portraying users as victims of Silicon Valley turns out not to be convincing. With Slavoj Žižek, we can say that we know social media is evil, but continue to use it. "What makes our situation so ominous, is the all pervasive sense of blockage. There is no clear way out, and the ruling elite is clearly losing its ability to rule."[7] Our environment and its operating conditions have been dramatically transformed, and yet our understanding of such dynamics lags behind. "The barbed wire remains invisible," as Evgeny Morozov once put it.

The problem has yet to be identified: there is no "social" anymore outside of social media. In Italian slang, social media has already been shortened: "Are you on social?" This is our Society of the Social.[8] We stare at the black box, wondering about the poverty of today's interior life. To overcome the deadlock, this book sets out to integrate a radical critique. It seeks alternatives by staging a subjective encounter with the multitude and their intimate dependencies on their mobile devices.

Internet culture is exhibiting signs of an existential midlife crisis. As Julia Kristeva once wrote: "There is nothing sadder than a dead God." The newness is gone, the innovation has slowed, the user base stabilized. In contrast to 1990s nostalgia, we can't really say there was ever a happy period of young adulthood. As in most non-Western cultures, it was straight into marriage at a young age with all the restrictions that come with it. Who dares to refer to "new" media anymore? Only innocent outsiders occasionally mention this once promising term. If anything, there seems to be a rapid spread of the retrograde, a yearning for the earlier and simpler days. What are we to make of this romantic nostalgia for the birth of virtual reality, the clumsy early web interfaces, and the net.art pioneers? Claude Levi-Strauss came up with a possible explanation: "Man never creates anything truly great except at the beginning: in whatever field it may be, only the first initiative is wholly valid. The succeeding ones are characterized by hesitation and regret, and try to recover, fragment by fragment, ground that has already been left behind."[9]

This volume, the sixth in my internet chronicles,[10] struggles with a digital realm that not only blends into the everyday, but increasingly impinges upon it—contracting our abilities and constraining our realities. This book deals with social media issues such as the selfie cult, meme politics, internet addiction and the new default of narcissist behavior. Two decades after dotcom mania we should be able to answer the question of how second order social media operate—but we can't. So while the social media question may be omnipresent, if we want to stand up against [insert your pathology here] by design we first have to understand its inner workings and operations unraveled here through the vector of distraction and sadness. The mechanisms of sadness are followed by a second section focused more on theory and strategy, from the "platform" concept to the invisibility of technological violence. The third section deals with the selfie craze, its anonymous mask design counterpart and whether progressive memes are possible in the first place. The final section examines the corporate data extraction industries and surveillance systems that orient mass behavior into a new form of social alienation. The concept of the commons runs counter to these logics, and I end by asking whether it offers a possible way out.

What happens when theory no longer presents itself as a grand design and is consumed as an afterthought? The internet is not a field in which public intellectuals play any role to speak of. Unlike previous eras, intellectual ambitions have to be modest. Before we design alternatives and formulate regulatory principles, it is vital to understand the psychology of social media platforms. *Sad by Design* combines radical internet critique with a confrontation of the all-too-real mental ups and downs of social media users. As Clifford Geertz observed, "understanding a people's culture exposes their normalness without reducing their particularity." For Geertz, "the study of culture penetrates into the very body of the object—that is, we begin with our own interpretations of what our informants are up to, or think they are up to, and then systematize those."[11] This book embraces Geertz's challenge, analyzing aspects of today's online cultures that many users experience, from feelings of emptiness, numbness and indifference through to the contradictory attitudes toward the selfie and the regressive politics of memes.

We seem disenchanted with our de-facto online cultures. British think tank Nesta neatly summed up our current condition. "As the dark side of the internet is becoming increasingly clear, public demand for more accountable, democratic, more human alternatives is growing." Yet the researchers are also honest enough to see that challenging the existing dynamics won't be easy. We are at an impasse. "The internet finds itself dominated by two ruling narratives: the American one, where power is concentrated in the hands of just a few big players, and a Chinese model, where government surveillance appears to be the leitmotif. Between Big Tech and government control, where does this leave citizens?" To label social media users as citizens is obviously a political framing, common lingo within NGO "global civil society" circles. Is this our only option to escape the consumer identity? Nesta put two strategic questions on the table: "Could Europe build the kind of alternatives that would put citizens back in the driver's seat?" And, rather than trying to build the next Google, should Europe focus on building the decentralized infrastructures that would prevent the next Google emerging in the first place?

The current state of the social should hardly be surprising. Technical media have long been socially antagonistic, undermining and isolating rather than connecting. In *Futurability*, Franco Berardi marks the late

1970s as the dividing line, the moment when social consciousness and techno-revolution diverged. This is when we "entered the age of techno-barbarism: innovation provoked precarity, richness created mass misery, solidarity became competition, the connected brain was uncoupled from the social body and the potency of knowledge was uncoupled from social welfare."[12] As Bernard Stiegler stated, the speed of technical development has continued to accelerate, "dramatically widening the distance between technical systems and social organization as if, negotiation between them appearing to be impossible, their final divorce seem inevitable."[13] For The Invisible Committee, social media "work towards the real isolation of everybody. By immobilizing bodies. By keeping everyone cloistered in their signifying bubble. The power play of cybernetic power is to give everyone the impression that they have access to the whole world when they are actually more and more separated, that they have more and more 'friends' when they are more and more autistic."[14]

What's to be done with social media? The last few years have been dominated by a profound confusion. For some, non-use seems to be a non-starter. Evgeny Morozov, for example, tweets: "I don't want #Zuckerberg to resign. And we don't need to #deleteFacebook: it's as realistic as saying #deleteroads. What we need is a New Deal for #data. #Europe has to wake up!" And, while Siva Vaidhyanathan criticizes Facebook fiercely, he refuses to leave and delete his account. For others, non-use is precisely the answer. An early proposal could be Ulises Mejias' *Off the Network, Disrupting the Digital World* from 2013, a book that claimed to "unthink the network logic."[15] More recently, but along these same lines, the "right to disconnect" movement has been starting to taking shape.[16] Take the offline-only Disconnect magazine, an anthology of commentary, fiction, and poetry that can only be read if you disable your WiFi.[17] Along with (grudging) use or non-use, a third approach might be filed under misuse. In a *Guardian* article titled "How to Disappear from the Internet", Simon Parkin provided (online) readers with a manual on how to become a digital ghost. "Deleting stuff is just useless" he asserted. His advice instead? Create fake accounts and misdirect searches. His conclusion, which makes his headline misleading at best, is that it is almost impossible to disappear. Options are limited to reputation management, either painstakingly conducted by ourselves— or for those with the money, carried out by specialized companies.

What if it's too late to leave Google, Twitter, Instagram or WhatsApp, no matter how digitally detoxed we become in other spheres of life? Let's face it, in the eyes of Silicon Valley, the offline, off-the-grid Burning Man experience once a year and the countless daily online Facebook visits are not opposites—they are complementary arrangements. Ergo, we are both offline and online.[18] Critique finds itself in a similar, contradictory position. The world has caught up with his arguments, Andrew Keen admits in his 2018 book *How to Fix the Future, Staying Human in the Digital Age*. Keen asks how we can reassert our agency over technology. We're not backseat drivers after all. As opposed to the protection of privacy, a demand that many consider Euro-centric and bourgeois, Keen instead demands data integrity. The fiddling with data has to stop. "Surveillance ultimately isn't a good business model. And if there's one thing that history teaches us, it's that bad business models eventually die."[19] He lists John Borthwick's "five bullets to fix the future: open technology platforms, anti-trust regulation, responsible human centric design, the preservation of public space and a new social security system."[20]

Yet the agency needed to implement these fixes seems hamstrung. Internet critics have limited power. Unable to network or to escape "old media", they have been pigeon holed into the role of the individual expert or commentator, excluded from any wider public dialogue about what's to be done. Academics too seem somewhat impotent. Driven by a logic of peer review and ranking, they publish inside the sealed universe of the journal with its limited access and even more limited impact. So while researchers certainly collect valuable evidence about the economic might of social media platforms, tech criticism at large remains scattered—incapable of institutionalizing its own practice and creating more cohesive schools of thought.

Recently we're witnessing the rise of peak data. Like peak oil, this is the theoretical point when the maximum rate of data extraction has been attained. From a user perspective, data are not consciously produced from intentional labor. Data collection becomes ubiquitous, an ever present procedure triggered by any movement, any act, any click or swipe. From a corporate perspective, data storage seems limitless; capacity is no longer a scarce resource. So although most (AI) pundits will tell you otherwise, the big data hype has reached its peak. Gartner, for example, had already dropped big data out of its hype cycle back in 2015. Peak data is the moment when the internet giants already know everything about you, the moment when additional details begin to tip the balance and cause their data regime to (slowly but inexorably) implode. This is the turning point. After this moment—and against the evangelists of eternal growth—each piece of data has the potential to make the entire collection less valuable, not more. After this moment, the value of extra data diminishes to a zero point, running the risk of polluting profiles in such a way that they disintegrate.

The data phantom of the self begins to crumble. The system produces such amounts of data that either everyone will become a suspect—or no one. Vital details will no longer be spotted. The production of information, once defined as the production of meaningful differences, is such that it flips and turns to zero: system overload. The goldmine of data suddenly becomes digital garbage. Companies like Google are aware of the dangers of such Hegelian turns and set out to rescue its valuable data assets.[21] It's worth remarking that such a policy shift does not come from any popular uprising against social depletion due to the takeover of intelligent machines. No, this is a strictly internal initiative aimed at self-preservation. In the new version of Android none of the tracking functionalities have been removed. Google simply collects less data—for its own well-being.

Platforms scramble to counter peak data by announcing new measures. For the first time, Google's Android operating system will be premised on restraint and reduction: "Instead of showing you all the ways you can use its phone operating system to do more, it's creating features to help you use it *less*."[22] The proposed dashboard will tell you "how often, when, and for how long you are using every app on your phone. It will also allow you to set limits on yourself." Think here of applying quantified-self dashboards like Fitbit to your phone's social media apps, making it easy to turn off notifications. "When bedtime arrives, your phone will automatically go into Do Not Disturb mode."

Other products follow suit. Google Search, for its part, responds to peak data with a new plan to show you "more useful ads". In a similar shift, the new update for Google's YouTube app includes a setting where the app reminds users to "take a break" from watching videos.[23] And in parallel to these moves, Google has launched a "wellbeing" campaign. The slogan? "Great technology should improve life, not distract from it."[24] Which values are emphasized when we progress toward a higher stage of development? Improved multi-tasking? This recent shift to self-limit is strange indeed. Will Google ultimately slow down real-time exchanges in order to build in reflection? What if improvement can only be achieved by speaking up against the dominant (and deadly) culture? Why should time-well-spent technology help you switch off?[25] Such responses to peak data are pre-emptive, striving to prevent disaster. With the danger of entropy looming from the (near) future, data collection is no longer an end in itself. For tech titans, the critical next step could be to press value out of the collected data whilst not upsetting users. This profile rescue plan is sold to the user as a contribution to their digital well-being, a gesture of "corporate responsibility". We could call this "backlash by design". Google has already anticipated any possible discontent. In a "precrime" *Minority Report* gesture, this response skips the resistance phase and installs the Hegelian synthesis preliminarily. We've overcome the culture of appropriation.

Silicon Valley already knows that we want to wind down. How will users respond to the default moralism of such changes? Against such do-gooder gestures, we should consider collectively implementing "data prevention" principles ourselves.

In the face of these conditions, we need internet studies more than ever. And yet somehow it has failed to be recognized and supported as a serious discipline. To modify Habermas, we can speak of the "unfinished project" of digitization as the latest stage of modernization, one that the post-1968 *Bildungselite* categorically blended out, convinced that the engineering buzz that produced internet tools would not affect them. Whereas we can still study film, theatre and literature, this is not the case with the internet, which has consistently failed to establish itself as a distinct academic discipline with its own full-time BA, MA and PhD programs. To defend this gap, institutions trot out the same line that "it's still early days"—as if not enough people are yet using the internet. Where is our "conflict of the faculties"? Worldwide, no one seems to be willing to take charge, to make that first (shaky but significant) step. Artistic new media programs have silently been closed down, have been merged into harmless, inward-looking academic enterprises such as "digital humanities", or have been subsumed into the broadcast logic of media and communications. As a result, the white male geeks from engineering and would-be venture capitalists from business schools have achieved cultural dominance—endlessly replicating Silicon Valley schemas and leaving those with a social science, arts and humanities or design background on the sidelines.

Italian Arabist and fellow activist Donatella della Ratta, who teaches digital culture at John Cabot University in Rome, adds another element:

The online subject is so deeply involved, that she can no longer see the phone, nor the internet. The young generation is not concerned with the technological device itself, they have simply erased it, forgotten it. My students are bored if I talk about technology per se. They want to talk about feelings, about their bodies and emotions… they simply don't see technology anymore.

What are the consequences of this rapidly spreading tech fatigue, right at a moment when controversies have finally reached the traditional political arena? As sociality becomes exhausted, decisions about commitment and connection become confused. "One has to know what to commit to and then commit to it. Even if it means making enemies. Or making friends. Once we know what we want, we're no longer alone, the world repopulates. Everywhere there are allies, closenesses, and an infinite gradation of possible friendships."[26] Contrast this ambitious, decisionist dream of The Invisible Committee with Mark Fisher's observation about the lack of self-motivation amongst students and a lack of sanctions if they're absent or do not perform well. "They typically respond to this freedom not by pursuing projects but by falling into hedonic (or anhedonic) lassitude: the soft narcosis, the comfort food oblivion of PlayStation, all-night TV and marijuana." Confronted with permanent information overload, millennials are said to be "overconfident", politely refusing to "learn more" and instead becoming attracted to "things that are more important". The concept of a social interior is no longer a paradox.

So before we rush headlong into debates about alternatives and strategies, *Sad by Design* feels the need to explore this rather vague, undefined realm of decision fatigue and ego depletion. This time there will be no travelogues, no lavish reports about Institute of Network Cultures initiatives such as Unlike Us, Video Vortex and MoneyLab. The market demands that I focus and present online despair in its most attractive form. In earlier essays I've written on nihilist blogging and the psycho-pathology of information overload. This book picks up those threads, examining in particular the interplay between our mental state and the technological condition. Here I investigate social reality from mental perspectives such as distraction and sadness. The book title can be read as a triangulation of addiction by design, the famous study on Las Vegas slot machines by Natasha Dow Schüll, the distraction by design of James Williams and the privacy by design of Ann Cavoukian.

Last but not least, let's look into the word "design" from this book's title. Is another design possible?[27] It is one thing to deconstruct the paint-by-numbers innovation of management-led design thinking. What role can design (and aesthetics in general) still play to overcome the current stagnation? One possible road here is to critically assess real existing design cultures, before we rush into the promotion of one radical design concept over another. We can't just have a life. We are condemned to design it. Benetton's colorful '90s photography of global misery has become a daily reality. Slums are flooded by designer clothes and footwear. Versace refugees are no longer rarities. Envy and competition have turned us into subjects of an aesthetic conspiracy that is impossible to escape. The McLuhanesque programmatic, "help beautify junk yards", is now a global reality. Gone are the days when Bauhaus design was supposed to empower the everyday reality of the working class. We're well past the point of design as an extra layer, an enhancement aiming to subtly assist eye and hand. Design is no longer a pedagogic discipline that intends to uplift the taste of the normies in order to give their daily lives sense and purpose. We're going for the lifestyle of the rich and famous. The ordinary is no longer enough; the mantra is onwards and upwards. We, the 99%, claim the exclusive lifestyle of the 1%. This is the aspiration of planet H&M.

Much like pre-torn and bleached denim, all our desired commodities have already been used, touched, altered, mixed, liked and shared before we purchase them. We're pre-consumed. With the radical distribution of funky lifestyles comes the loss of semiology. There is no control anymore of meaning. Brands can mean anything for anyone. This is precarity of the sign.

Our beautified mess is no longer an accident or a tragic sign of a never-ending decay, but rather an integral part of the overall layout. Today's design culture is an expression of our intensely prototypical lives. We are the experience junkies who desire to wring out life's pleasures, to thoroughly exhaust it. And yet it's remarkable how little transformative progress we have made. We want so much, and make so little. Our precarious state has become perpetual.

When we confront ourselves with sci-fi precarity—that strange techno-reality ahead of us—the first association that comes to mind is the conformist 1950s. Sure, we wished we lived in a *Blade Runner* movie, but our reality more closely resembles a Victor Hugo novel or a Douglas Sirk film in which the hyper-real takes command. Boredom, anxiety and despair are the unfortunate default. That's "real existing precarity", comparable to "real existing socialism" in the outgoing Cold War period. Casual precarity, everywhere you look. The terror of comfort drives us mad. The flatness of it all is contrasted and accelerated by the occasional modernist IKEA style that, in theory, should cheer us up, but in the end only provokes an inner revolt against this manufactured reality. What's to be done with workers that have nothing to lose but their Ray-Ban sunglasses? We can't wait for Godot, not even for a split second. No matter how desperate the situation, the uprising simply won't happen. At best we attend a festival, expand our mind and body—and then sink back into the void.

Once the silence has faded away, we step out of our arty-geeky-academic echo chambers. The current political situation demands we refrain from techno-solutionist proposals and instead migrate these supposedly narrow internet issues into larger contexts such as precarity, post-colonial tech politics, gender issues, climate change action or alternative urbanism. Despite all the potential for fatalism and introspection, let's stick to Mark Fisher's slogan: "pessimism of the emotions, optimism of the act."[28] As a gesture to this moment, my investigation into critical internet cultures concludes with a contribution to the commons debate. As Noam Chomsky said:

> There is a great deal that we can do to bend the arc of history towards justice, to borrow the phrase that Martin Luther King made famous. The easy way is to succumb to despair, and help ensure that the worst will happen. The sensible and courageous way is to join those who are working for a better world, using the ample opportunities available.[29]

1

Overcoming the Disillusioned Internet

Brand tagline: Properly Distracted, Totally Extracted"—"Artificial intelligence is not the answer to organized stupidity." Johan Sjerpstra—"Please don't email me unless you're going to pay me." Molly Soda—"Late capitalism is like your love life: it looks a lot less bleak through an Instagram filter." Laurie Penny—"Wonder how many people going on about the necessity of free speech and rational debate have blocked and muted trolls?" Nick Srnicek—"Post-truth is to digital capitalism what pollution is to fossil capitalism—a by-product of operations." Evgeny Morozov—"I have seen the troll army and it is us." Erin Gün Sirer.

Disenchantment with the internet is a fact.[1] Enlightenment does not bring us liberation, but depression. The once fabulous aura that surrounded our beloved apps, blogs, and social media has deflated. Swiping, sharing, and liking feel like soulless routines, empty gestures. We've started to unfriend and unfollow, yet we can't afford to delete our accounts since this implies social suicide. If "truth is whatever produces the most eyeballs," as Evgeny Morozov states, a general click strike seems the only option left. As this is not happening, we feel trapped, consoling ourselves with memes.

The multi-truth approach of identity politics, according to Slavoj Žižek, has produced a culture of relativism.[2] Lippmann and Chomsky's "manufacturing consent" has come to a halt. As Žižek explains in a British TV interview, the Big Other has vanished.[3] There is no BBC World Service anymore, the moderate radio voice that once provided us with balanced opinions and reliable information. Every piece of information carries the suspicion of self-promotion in it, crafted by public relation managers and spin doctors—and ourselves as users (we are our own marketing intern). What's collapsing is the imagination of a better life. In protesting, it is not the wretched of the earth who are rebelling because they've got nothing left to lose, but rather the stagnating middle class and young urban professionals who are facing permanent precarity.

Mass conformity didn't pay off. Once the love affair with the app is over and the addiction reveals itself, the mood flips to loathing and thoughts turn to quitting cold turkey. What comes after the Exorbitant Detriment? After hubris comes guilt, shame, and remorse. The question is how the current discontent will ultimately play out on the level of the architecture of the internet. What is techno-repentance? How can we reintroduce the decentralized web, the enlightened American tech community asks itself, after decades of uncritical support for their own addictive monopolies-in-the-making.[4] Their answer? Write more code. In contrast, the European response to the "broken internet" is a public infrastructure initiative called technological sovereignty and the "public stack" (more on this in Chapter 5).

What some see as a relief is experienced by many as frustration, if not hatred. The online Other cannot possibly be classified any longer as a Friend. "If people in the outside world scare you, people on the internet will downright terrify you" is a general warning that is applicable to all sites. Troll awareness has never been this high. Unable to escape and condemned to remain online, our existential encounter with the troll seems inevitable. Users are under threat of socioeconomic collapse and, once poor, being subjected to the post-money economy, in which only imaginary entities circulate. Once having been written off, being online is their last refuge.

"We're terrofucked." That's how Jarett Kobek summed up the general feeling in his 2016 *I Hate the Internet* novel. The guilt and frustration is both personal and political, on a global scale. Set in the gentrified streets of San Francisco, the story describes how computers coordinate the exploitation of "the surplus population into perpetual servants." What happens once the realization sinks in that "all the world's computers were built by slaves in China" and that you are the person who uses those same devices? What happens when we're personally addressed as the guilty partners, "suffering the moral outrage of a hypocritical writer who has profited from the spoils of slavery"?[5]

What if the current internet economy of the free is the default future scenario for the 99%? This is the intriguing part of Kobek's DIY philosophy, which he presents as a science fiction of the present. What will happen when the concentration of power and money in the hands of the few indeed becomes irreversible and we abandon all hope of revenues being redistributed? For Kobek, this is already the case. Traditional money has failed, replaced by the micro-fame of influencers, "the world's last valid currency" which is even more susceptible to oscillations than money. "Traditional money had ceased to be about an exchange of humiliation for food and shelter. Traditional money had become the equivalent of a fantasy world."

Kobek profiles himself as a proponent of the "bad novel," in contrast to the CIA-sponsored literary fiction of the Cold War, called the "good novel"—a category that continues to exist in the work of Jonathan Franzen, "who wrote about people from the American Middle West without much eumelanin in their epidermises." Bad novels are defined here as stories that "mimic the computer network in its obsession with junk media, in its irrelevant and jagged presentation of content," stories filled with characters who have a "deep affection for juvenile literature" such as Heinlein, Tolkien, and Rand. This all makes you wonder in which category Dave Eggers's update on *1984*, *The Circle*, might fit. Can this story about the prediction economy, enforced by a fictitious merger of Google and Facebook, be classified as the ur-bad novel in this category? What happens when we are no longer able to distinguish between utopia and dystopia?

For Kobek, fame and the internet are both devices for stripping us of agency. The promise of fame deludes people with images of grotesque success. As long as they believe in their dreams, everyone is a performer and a celebrity, emulating examples like Beyoncé and Rihanna, who are inspirations rather than vultures. Such celebrity cases showed "how powerless people demonstrated their supplication before their masters." Fans are fellow travelers on a journey through life; they are not consumers who purchase a product or service. According to Kobek, "the poor are doomed to the Internet, a wonderful resource for watching shitty television and experiencing angst about other people's salaries." Built by "pointless men," the internet invokes nothing but trash and hate, leaving the poor empty-handed, with nothing to sell. The poor make money for Facebook; it will never be the other way round.

Kobek's style has been compared with Houellebecq's due to the harshness of their characters. We wander through the cynical startup environment of "throwing stones at the Google bus" Bay area, yet Kobek refuses to take us inside. This is the perspective of the outcast and the desperate, a perspective that at least promises some real insights. We already note the desert-like collective imaginary of the geek class, that mixture of Hacker News, Reddit, 4Chan, games and porn. Unlike a cyberpunk novel, we don't enter cyberspace, we don't plug in and swipe through the profiles flowing through Instagram. This is not about an illusion of the end. And this is the main difference from the revolutionary-utopian 1968 generation: we have the uncanny feeling that something has barely started. In this dystopian, hyper-conservative era, we no longer face up to the historical duty to confront the finality of society's episodes such as the welfare state, neo-liberalism, globalization, or the European Union. Instead we've been lured into a perpetual state of retromania, because, as the late Mark Fisher pointed out, it is the present that went missing ("Make America Dank Again").

Inside these pseudo-events there is no chronology, no development, no beginning or middle, let alone an end. We're beyond the terminal process, *jenseits* the postmodern patchwork. Everything is accelerating. This must be the catastrophic twenty-first-century style featured in so many films. Still we remain encapsulated, captured within cybernetic loops that go nowhere, watching transfixed as meaningless cycles of events, series, and seasons pass by. What happens when the anxiety of information saturation flips to become a profound feeling of emptiness? Once we've passed this point, the digital neither disappears, nor does it end. Events simply no longer turn into Roman spectacles. We instead experience the simulacrum as prime reality. We cannot process such a sudden overproduction of reality.

We no longer turn on the television news thinking that we're watching a film. We've moved on. It is not life that has become cinematographic; it is the film scenario and its effects that shape the grand designs of our technological societies. Films anticipated our condition, and now we're situated in the midst of yesteryears' science fiction. *Minority Report* is now a techno-bureaucratic reality, driven by the integration of once-separate data streams. *Black Mirror* is not a joke. Virtual reality actually feels like *The Matrix*. Trump's reality TV shows proved to be rehearsals. His tweets are actually US policy. All this makes us long for truly untimely, weird fiction. The avant-garde logic still seems alive with the role of bohemian artists taken over by engineers and entrepreneurs. We've left behind the stage of art and entertainment as "proposals" and "scenarios". The last industry to deal with the whirlpool of the fake and the real is the news industry itself. Hyperreality becomes our everyday situation—regardless of whether you perceive it as boring or fringe.

Let's look at radical disillusion as form and celebrate the return of its high priest, Jean Baudrillard. Social media rage is not just a medical condition of the few; it is the human condition. Will the disenchantment turn into a revolt, as Camus once contemplated? The spiritual exhaustion is certainly there (#sleepnomore). Empty-handed, we discuss one brilliant yet powerless critique of the algorithm after another. To put it in spatial terms, the infinite world of cyberspace—a room containing a house containing a city—has collapsed into an arid, exposed landscape in which transparency rapidly shades into paranoia. Rather than lost in a labyrinth, we're thrown out into the open, watched, and manipulated, with no command centers in sight.

The *mille plateaux* of tweets, blogs, Instagram stories, and Facebook updates have created a culture of deep confusion. Fragmentation was supposed to enrich us. Remember radical difference as fractal beauty? All good. Then why should we now have to pay the bill for all the unforeseen consequences? This wasn't supposed to happen. Is this the Derridaland we once dreamed of? Mainstream media play a decisive role in this process of decay. Mainstream media's role as "clearing houses" for facts and opinions has been undermined for decades by growing centripetal forces in society that no longer accept particular baby-boom sentiments such as truth and independence. Yet while their legitimacy has faded, their influence remains substantial. This creates an atmosphere of permanent ambivalence.

After decades of hard work to deconstruct the dominant ideology of the mainstream media, there is no way back. The liberal consensus is broken. The stunning inability of "the press" to deal with the changes in society (from climate change to income inequality) has lead to a widespread form of indifference. Why bother with the living dead? The theoretical blind spots of the successive postmodern generations are too numerous to list. The big elephant in the room here is Jürgen Habermas. Many of us still subscribe to his notion of the bourgeois public sphere as an arena where different opinions compete in a rational dialogue— even if we do not believe in the core values of Western society such as democracy. And who would even be the counter public in this context? The "user generated content" of 4Chan, Reddit or YouTube channels such as PewDiePie? What's the organized answer to all this? Moral condemnation and denial. And ourselves as activists, what do we have to offer? What does a contemporary version of Indymedia look like? And where is it, now that we need it so badly? If a federated model of bottom-up news filtering is possible, then let's build it.

There is a crisis of "participatory culture". Let's look at the example of danah boyd and how she's deconstructing the media literacy discourse about which so many of us had high hopes. A cynical reading of the news has overshadowed critical capacities. In the aftermath of Donald Trump's 2016 election, she asked if media literacy has backfired.[6] It's lazy to only blame trolling, clickbait, and fake news for the drop in news legitimacy. For the pre-internet baby-boom generation, literacy was synonymous with the ability to question sources, deconstruct opinions, and read ideology into quasi-neutral messages. Today the meaning of literacy has shifted, referring to the ability of citizens to produce one's own content in the form of responses, contributions, blog postings, social media updates and images uploaded to video channels and photo-sharing sites.

However, this shift from critical consumer to critical producer came with a price: information inflation. The authority to filter news shifted from top-down broadcasting media to tech giants. According to boyd, media literacy has come to resemble a distrust of media sources, and no longer fact-based critique. Instead of considering the evidence of experts, it has become sufficient to bring up one's own experience. Outrage has triumphed, reasonable debate atrophied. The result is a highly polarized culture that favors tribalism and self-segregation.

The current situation demands a rethinking of the usual demands of activists and civil society players with regard to "media literacy". How can the general audience be better informed? Is this an accurate diagnosis of the current problem in the first place? How can holes be made in filter bubbles? How can Do-It-Yourself be a viable alternative when social media are already experienced in such terms? And can we still rely on the emancipatory potential of talking back to the media via the familiar social networking apps? How does manipulation function today? Is it still productive to deconstruct *The New York Times* (and its equivalents)? If the seventies produced *How to Read Donald Duck*[7] who's going to write *How to Use Facebook*? How would you explain the workings of the Facebook newsfeed to its user base? Is it still a black box?

If we want to blame algorithms, how can we popularize their complexity to large audiences? A case in point might be Cathy O'Neil's *Weapons of Math Destruction*, in which she describes how "ill-conceived mathematical models micromanage the economy, from advertising to prisons."[8] Her question is how to tame and, yes, disarm dangerous algorithms. Such mathematical models are not neutral tools. However in everyday life, we increasingly experience ranking as destiny. "Promising efficiency and fairness, they distort higher education, drive-up debt, spur mass incarceration, pummel the poor at nearly every juncture, and undermine democracy." In this account of her jobs in numerous industries, she shows that this software is "not just constructed from data but from the choices we make about which data to pay attention to—and which to leave out. Those choices are not just about logistics, profits, and efficiency. They are fundamentally moral." And class-biased, she adds: "The privileged are processed by people, the masses by machines." Once installed and running for a while, these "difference engines" create their own reality and justify their own results, a model that O'Neil denotes as self-perpetuating and highly destructive.

Techniques such as leaks, fake news, social bots, kompromat, and agitprop confuse the political climate. Disorientation is sufficient; it is no longer necessary to manipulate election outcomes. In this post-factual era we are left with the instant beliefs of celebrity commentators and media experts. Look at Donald Trump's tweets, an ultimate form of media literacy and a perverse epiphany of self-expression.[9] Personal tweets have become indistinguishable from policy, state propaganda, and info warfare. In this sense, power no longer operates through the seduction of pornographic overexposure to high-res 3D images. This is not Big, but rather Singular Data—tiny messages with a "tremendous" fall-out. At this level, we leave behind the realms of both Hollywood glamor and reality TV and enter the real-time realm of communication-with-consequences, a next-level hybrid in which sovereign executive power and marketing become inseparable.

No one exemplifies this power-as-performance better than Trump. This is a man who "seems supremely cognizant of the fact that he is always acting. He moves through life like a man who knows he is always being observed." This pre-election quote is taken from a June 2016 piece in *The Atlantic* called "The Mind of Donald Trump."[10] There he is described as "lummoxing": "sky-high extroversion combined with off-the-chart low agreeableness." The list of his characteristics is breath-taking. He's described as a dynamo—driven, restless, unable to keep still, getting by with very little sleep. A cardinal feature of high extro-version is his relentless reward seeking. Prompted by the activity of dopamine circuits in the brain, highly extroverted actors are driven to pursue positive emotional experiences. "Anger can fuel malice, but it can also motivate social dominance, stoking a desire to win the adoration of others… Anger lies at the heart of Trump's charisma… dominated by "the ebullient extroversion, the relentless showmanship, and the larger-than-life celebrity," who never thinks twice about the collateral damage he will leave behind.

Highly narcissistic people draw attention to themselves. Repeated and inordinate self-reference is a distinguishing feature of their personality. Over time, people become annoyed, if not infuriated, by their self-centeredness. When narcissists begin to disappoint those whom they once dazzled, their descent can be especially precipitous. There is still truth today in the ancient proverb: "pride goeth before the fall." The world is saturated with a sense of danger and a need for toughness. The world cannot be trusted. In this brutal arena, the successful hero is the ferocious combatant who fights to win. Are you preoccupied with fantasies that the world is ending because of the selfishness of others? "Who, really, is Donald Trump? What's behind the actor's mask? I can discern little more than narcissistic motivations and a complementary personal narrative about winning at any cost. It is as if Trump has invested so much of himself in developing and refining his socially dominant role that he has nothing left over to create a meaningful story for his life, or for the nation. *It is always Donald Trump playing Donald Trump*, fighting to win, but never knowing why."[11]

Where might we find the starting point for today's philosophy of disbelief? Should we look for a secular follow-up to the critique of religion? What is atheism within the context of information? What's our totem and what's taboo? The multiplicity of sources and points of view, once celebrated as "diversity of opinion", is now reaching its nihilistic "zero point". Rather than an accumulation of meaning leading to potentially critical insights (or even to knowledge), information implodes into a pool of indifference (a situation possibly leading to the disappearance of channels such as Twitter, since they thrive on individual expressions, judgments, and preferences).

These days, institutional dogmas are hidden inside media folklore, hardwired into network architectures, steered by algorithms. The mental rejection of authority is now so widespread and has sunk so deep into daily routines and mentalities that it has become irrelevant whether we deny, endorse, or deconstruct a particular piece of information. This is the tricky aspect of the current social media disposition.

Barricaded in their bedrooms, meme producers have become immune to any criticism by third-way liberal moralists. Their firewall of indifference has not yet been hacked. Ironic deconstruction isn't doing the job either. Tara Burton says: "Given the ideological anarchy inherent in shitposting, it tends to defy analysis. Shitposters, who are bound by nothing, set a rhetorical trap for their enemies, who tend to be bound by having an actual point."[12] "Shitposting can't be refuted; it can only be repeated," Burton concludes. This is not the age of the Renaissance (Wo) Man; the disillusion is simply overwhelming.

We're overwhelmed by media events unfolding in real-time. Is this spectacle a smoke screen for more drastic, long-term measures? What's our plan? The politically correct strategies of civil society are all well-meant and related to important issues, but seem to be moving toward a parallel universe, unable to respond to the cynical meme design that is rapidly taking over key power positions. Are there ways to not just hit back, but also be one step ahead? What's on our mind?

How can we move from data to Dada and become a twenty-first-century avant-garde, one that truly understands the technological imperative and shows that *we* are the social in social media? In short, how do we develop, and then scale up, critical concepts, and bring together politics and aesthetics in a way that speaks to the online millions? Let's identify the hurdles, knowing that it's time to act. We know that making fun of the petty world of xenophobes isn't working. They're on the offense, not us. What can we do other than come together?[13] Can we expect anything from the designer as lone wolf? How can contemporary political labor be organized outside of Facebook and Twitter? Do we need even more tools to bring us together? Have you already started using DuckDuckGo, Meetup, Diaspora, Mastodon, DemocracyOS, or Loomio? Where are the collective dating sites for political activism? How can we design and then mobilize a collective, networked desire that unites us in a "deep diversity"? Is the promise of open, distributed networks going to do the job or are you looking for strong ties—and all the consequences that come with them?

Generations have studied the fatal mistakes made in the interwar period, but what's the conclusion now that we're moving toward a similar situation? What would a non-fascist life these days look like? Can we still be inspired by Hannah Arendt's *The Origins of Totalitarianism* (in which we find David Rousset's quote: "Normal men do not know that everything is possible"), Wilhelm Reich's *Mass Psychology of Fascism*, Adorno and Horkheimer's *Dialectics of Enlightenment*, Elias Canetti's *Crowd and Power*, and the opus that defined my own intellectual destiny, Klaus Theweleit's *Male Fantasies*? Needless to say, this is a subjective list, since there are so many classics in this genre.[14] Will these authors assist us in finding out what the defining causes of regression are? How can we locate the key issues, and then act upon them, knowing that we've identified the socio-psychological factors that are causing the tipping point?

2

Social Media as Ideology

"What does the money machine eat? It eats youth, spontaneity, life, beauty, and, above all, it eats creativity. It eats quality and shits quantity."—William S. Burroughs—"In data we trust." Priconomics—"The Internet fails to scale gracefully." Chris Ellis—"I want to be surprised by my own bot"— "There is a crack in everything. That's how the light gets in." Leonard Cohen—"Just did my sheepish biannual LinkedIn visit, which felt too much like my sheepish biannual sweeping of dry cleaner hangers into the bin." Dayo Olopade—Organic Reach Technologies (company)— "It's not a pilot study. It's small batch artisanal data." @AcademicsSay—"No Reply" The Beatles—"A Facebook-Op occurs when one takes a photo just to upload it to Facebook later." Urban Dictionary—"If you start to think that people are awful, you can always sign on to Twitter. Get some further proof. Then go on about your day." Nein—"The right people can work around a bad technology, but the wrong people will mess up even a good one." Kentaro Toyama—The thing you are supposed to be decentralizing is power." Sarah Jamie Levis—"You can wake up now, the universe has ended." Jim Stark

The Internet has reached its hegemonic stage. There is no longer a need to investigate the potential of "new media" and deconstruct their intentions. In the past decades it was premature to associate intensive 24/7 usage of the millions to deep structures such as the (sub)conscious. Now that we live fully in social media times, it has become pertinent to do precisely that: link *techne* with psyche. Contradictory consciousness management has superseded social anxiety about Bad Faith.[1] This has long been the thesis of Slavoj Žižek. Let's work on this thesis, taking seriously the cynical statement "They know what they do, but they do it anyway" and applying it to social media.

The effects of Edward Snowden's revelations have suffused deep into our daily surf and swap routines. We know we're watched by surveillance systems, but who can honestly claim to be aware of them? Artistic masks are promoted as protective shields of the face, but who actually wears them? The internet may be broken, as the phrase says (and IT engineers have reached a consensus about this troubling analysis), but this cannot be said of social media.[2] The same is true of Sherry Turkle's evidence that smart phones reduce our abilities to develop empathy and to enjoy solitude free of connected devices.[3] How hard has it become to confront offline boredom and to simply dwell in the spiritual act of "radical presence"? Admit it, it's straight out torture.

"You are what you share."[4] This slogan expresses the transformation of the autonomous unit of the self into an outwardly facing entity that is constantly reproducing its social capital by exposing value (data) to others. Let's face it: we refuse to perceive ourselves as slaves of the machine. The current platforms are scraping the social, but we politely reject experiencing it in this way. What does it mean when we all agree there is an addictive element to today's social media use, yet none of us is apparently addicted? Are we really returning only sporadically?[5] What exactly is being captured here? If anything, we're encapsulated by the social sphere as such, not by software, protocols, network architectures or the all too infantile interfaces.

Hypnotized by the spell of the social and led by the views and opinions of our immediate social circle, these are your daily routines: view recent stories first, fine-tune filter preferences, jump to first unread, update your life with events, clear and refresh all, mark as "not now", save links for later, see full conversation, mute your ex, setup a secret board, run a poll, comment through a social plug-in, add video to your profile, select a reaction (love, haha, wow, sad or angry), engage with those who mention you, track the changing relationship status of others, follow a key opinion leader, receive notifications, create a slideshow that links to your avatar, repost a photo, get lost in the double-barrel river of the timeline, block friends from seeing updates, customize your cover image, create some "must-click" headlines, chat with a friend, and notice that "1,326,595 people like this topic." Social media demands a never-ending show—and we are the performers. Always signed in, we keep circling back for more, until the #DigitalDetox app switches us off or we're summoned to enter different realms.

Social networking has expanded far beyond being a dominant discourse. Media here is not constrained to text and images, but comprises the operations of software, interfaces and networks, underpinned by technical infrastructures of offices and data centers, consultants and cleaners, which works intimately with the movements and habits of the connected billions. Overwhelmed by this complexity, internet studies has downgraded its attention from utopian promises, impulses and critiques to mapping the network's impact. From digital humanities to data science, we see a shift in network-oriented inquiry from Whether and Why, What and Who, to merely How—from a sociality of causes to a sociality of net effects. A new generation of humanistic researchers is lured into the big data trap, kept busy capturing user behavior whilst producing seductive eye candy for an image-hungry audience (and vice versa).

Without noticing it, we have arrived in the next, as yet unnamed stage, the hegemonic era of social media platforms as ideology. Of course, products and services are usually subject to ideology. We have learned to "read" ideology into them. But at what point can we convincingly say they have become ideology themselves? It is one thing to state that Facebook's CEO Mark Zuckerberg is an ideologue, working in the service of US-intelligence, or to document community or political groups using his platform in ways unanticipated by its original design. It is quite another to work on a comprehensive social media theory. Now is the crucial time for critical theory to reclaim lost territory and bring on exactly this: a shift from quantitative statistics and mapping to the messier, more subjective, but altogether more profound qualitative effects—the incomputable impacts of this ubiquitous formatting of the social. It is liberating for research to sever itself from the instrumental approach of (viral) marketing and public relations. Stop pandering and promoting, start analyzing and criticizing. Network technologies are rapidly becoming the new normal, withdrawing their operations and governance from view. We need to politicize the New Electricity, the privately owned utilities of our century, before they disappear into the background.

Now, a decade after the 2008 wave of internet criticism, the phase that featured Nicholas Carr, Sherry Turkle, Jaron Lanier and Andrew Keen is coming to a close. The easy opposition of Californian utopians vs. Euro pessimists has been superseded by larger planetary issues such as the future of work, climate change and political backlashes. The social, political, and economic promise of the internet as a decentralized network of networks lies in tatters. Social media alternatives, introduced during the turbulent year of 2011, haven't made much progress at all.[6] Further, despite all the well-meant critical predictions, the herds have not moved on to greener pastures elsewhere. The overall picture is one of stagnation in a field defined by the corporate domination of a handful of players. We are all stuck in the social media mud, and it's time to ask why.

Comparable to the late 1970s stagnation in mainstream media critique, a political economy approach will not be sufficient if we want to come up with workable strategies. We need to take internet critique beyond the normative regulation of behavior and politicize the anxiety of the youth and their particular addictions and distractions. How can we ground the critique in disciplines such as urban, post-colonial and gender studies and take over the digital realm from such corners? One possible way out could be a post-Freudian answer to the question: What's on a User's Mind?[7] We need to answer the question as to what social media actually offer. Which desires do they appeal to? Why is updating a profile such a boring, yet strangely seductive habit? Can we develop a set of critical concepts describing our compulsive attraction to social media without reducing it to the rhetoric of addiction?

The prominence of ideology as a central term in debates has faded away since the mid-1980s. The backdrop of ideology theory in the 1970s was the spectacular peaking of the power of the state apparatus (also called the welfare state) that was commissioned to administrate the post-war class compromise. Whilst Daniel Bell's *End of Ideology*, as proclaimed in 1960, had announced the victory of neo-liberalism at the end of the Cold War, there was an intuitive feeling that ideology (lowercase i) had not yet left the stage. Despite concerted efforts to diminish the role of public intellectuals and critical discourses, the World Without Ideas was not yet within reach.

The "Californian ideology" as defined in 1995 by Richard Barbrook and Andy Cameron helped us to trace the internet motives back to their Cold War roots (and the ambivalent hippie culture), as did Fred Turner's 2006 classic, *From Counterculture to Cyberculture*. But the historical perspective is not much use if it cannot explain social media's persistent success since the 1990s and its allure today. Now, as in the 1970s, the role of ideology in navigating the limits of existing systems is all too real. To study ideology is to take a closer look at this everyday life, here and now. What remains particularly unexplained is the apparent paradox between the hyper-individualized subject and the herd mentality of the social. What's wrong with the social? Better yet, what's right with it? Social positivity is as residual in California as it is in the Italian cyberspace scene, where a Gramscian embrace of the social network is even taken as a sign that the multitude can beat the mainstream in its own act of mediation. In this sense, Italian critics, activists and artists are not unlike many others—hyper-aware of all the controversies surrounding Silicon Valley services, and yet persistently positive about the magic potion called "social networking".

One function of ideology as defined by Louis Althusser is recognition—the (in)famous interpellation of the subject who is called upon. Building on this idea, we might speak of the process of becoming-user This is the unnoticed part of the social media saga. The platforms present themselves as self-evident—they just are. After all, they facilitate our feature-rich lives and everyone that counts is there. But before entering, everyone must first create an account, filling out a profile and choosing a username and password. Minutes later, you're part of the game and start sharing, creating, playing, as if it had always been like that. The profile is the a-priori, a component that the profiling and targeted advertising cannot operate without. It is through the gateway of the profile that we become its subject.

For Althusser we live inside ideology in this way—the formula applies in particular to social media in which subjects are addressed as users that do not exist without a profile. Though slightly authoritarian and hermetic, the use of ideology as a concept can be justified because social media itself is a highly centralized, top-down structure. In this age of platform capitalism, social media architecture actively closes down possibilities, leaving zero space for users to reprogram their communication spaces.

Despite all the postmodernism and cynical neo-liberalism that has deemed it redundant, the fact that ideology again rules is no surprise (in fact it is more remarkable how total the concept's fall from grace has been). The main issue is that we are less and less aware of how. Furthermore, when it comes to social media, we have an enlightened false consciousness—we know very well what we are doing when we are fully sucked in, but we do it anyway. This even counts at a meta-level for the popularity of Žižek's insights and could be one of the best explanations of his success. We're all aware of the algorithmic manipulations of the Facebook newsfeed, the filter bubble effect in apps and the persuasive presence of personalized advertisement. We pull in updates, 24/7, in a real-time global economy of interdependencies, having been taught to read the news feeds as interpersonal indicators of the planetary condition. So how does Louis Althusser need updating?[8]

Four decades after the Althusser era, we do not associate ideology with the state in the same way he and his followers did then. To qualify Facebook and Google as falling within the Althusserian definition of "ideological state apparatus" sounds odd, if not exotic. In this era of late neo-liberalism and right-wing populism, ideology is associated with the market, not with the state, which has withdrawn into the sphere of market security. But lest we forget, it was ideology theory itself that contributed to the crisis of Marxism. It opened up the various issues raised by the student, feminist and other new social movements, aggravating the stagnation and eventual bankruptcy of the Soviet Union. The growing interest in media and cultural studies did the rest.

Broadcast live to the world via satellite, the fall of the Berlin Wall in 1989 became an instant and unmanageable news item, hurled into circulation alongside every other story. Already then, weakened communist parties could no longer annex and contain the rainbow of justice and redistribution issues of the properly (or revolutionary) social state, let alone its counter-cultural practices. Because of this, the tactics of over-determination in the name of the working class also no longer worked. The so-called patchwork of minorities who refused the new normal were literally left to their own devices, devoid of any overarching political framework, let alone an organizational structure or even an antagonist. Within a decade, Marxist theory as ideology critique had lost the dominance of two of its defining centripetal forces: State and Party. As a result, ideology as a primary focus of attention in philosophy and social sciences largely disappeared. And this absence manifested in the common belief that while ideas still mattered, they were no longer able to rule people's lives. Today, ideas are praised because they can shape the future, but formalized into rules and norms, they are believed to be too rigid and static to rule over our messy, contradictory everyday lives under capital.

What is crusted as orthodoxy in Althusser can be updated through Wendy Chun's 2004 essay on software as ideology. Chun's work, along with Jodi Dean and others, spoke strongly to the media theorist coming to terms with the peak of neo-liberal transition and the triumph of proprietary software. 2004 was the golden era of Web 2.0, an era in which software was considered synonymous with, or even confused with, PCs and laptops. She wrote then: "Software is a functional analogue to ideology. In a formal sense, computers understood as comprising software and hardware are ideology machines." Chun observed that software "fulfills almost every formal definition of ideology we have, from ideology as false consciousness to Louis Althusser's definition of ideology as a 'representation' of the imaginary relation of individuals to their real conditions of existence."[9] In an age of embedded micro-perceptual effects and stream programming, ideology does not merely refer to an abstract sphere where the battle of ideas is being fought out. Instead, think more in terms of a Spinozian sense of embodiment—from the repetitive strains of swiping to the "txt neck" from peering down and the shoulders perpetually hunched over the laptop syndrome.

So Althusser needs adapting—and not just in terms of a class analysis. But it is remarkable how smoothly an Althusserian ideological framework still fits today's world. As Chun asserts:

> Software, or perhaps more precisely operating systems, offer us an imaginary relationship to our hardware: they do not represent transistors but rather desktops and recycling bins. Software produces users. Without operating system (OS) there would be no access to hardware; without OS no actions, no practices, and thus no user. Each OS, through its advertisements, interpellates a "user": calls it and offers it a name or image with which to identify.

We could say that social media performs the same function and is even more powerful.

Understanding social media as ideology means observing how it binds together the media, culture and identity complexes into an ever-growing cultural performance, tying together gender, lifestyle, fashion, brands, and celebrity gossip with news from the radio, television, magazines and the web—and recognizing that all of this is infused with the entrepreneurial values of venture capital and start-up culture, values that carry with them a shadow side of declining livelihoods and growing inequality.

"What are you doing?" said Twitter's original phrase. The question marks the material roots of social media. Social media platforms have never asked what you are thinking (or dreaming for that matter). Twentieth-century libraries are filled with novels, diaries, comic strips, and films of people expressing what they were thinking. Yet in the age of social media, we seem to confess less of what we think. It's considered too risky, too private. We share what we do and see, but always in a staged manner. We share judgments and opinions, but no thoughts. Our Self is simply too busy for that. Flexible, open, sportive, and sexy, we are always on the move, always ready to connect and express.

With 24/7 social visibility, apparatus and application become interiorized in the body. This is a reversal of Marshall McLuhan's Extensions of Man into an Inversion of Man. Once technology entangles our senses and gets under our skin, distance collapses and we no longer feel that we are bridging distances. With Jean Baudrillard, we could speak of an implosion of the social into the hand-held device in which an unprecedented accumulation of storage capacity, computational power, software and social capital is crystallized. Steered by our autonomous fingertips, things get thrust in our face and poured into our ears. This is what Michel Serres admires so much in the navigational plasticity of the mobile generation: the smoothness of their gestures, symbolized in the speed of the thumb, that are able to send updates in seconds, master the microconversation, and grasp the mood of a global tribe in an instant.

To stay within the French realm of references, social media as an apparatus of sexy and sportive active acting makes it a perfect vehicle for the literature of despair, epitomized in Michel Houellebecq's messy body (-politics): "Our civilization suffers from vital exhaustion," he writes in *Whatever,*

> In the century of Louis XIV, when the appetite for living was great, official culture placed the accent on the negation of pleasure and of the flesh; repeated insistently that mundane life can offer only imperfect joys, that the only true source of happiness was in God. Such a discourse would no longer be tolerated today. We need adventure and eroticism because we need to hear that life is marvelous and exciting.[10]

There is a self-evident quality to social media. Swiping and tapping through updates, users surround themselves with an illusion which feels natural from the very first time. There is no steep learning curve or rite of passage, no blood, sweat and tears needed to break into the social hierarchy. From day one the network configuration makes us feel at home. It is as if Messenger, WhatsApp, WeChat and Telegram have always existed. But it is precisely this immediate and effortless familiarity that becomes the main source of discontent down the track. We're no longer playing, like in the good old days of Lamda MOO or Second Life. Intuitively we sense that social media is an arena of struggle where we display our "experientalism",[11] where hierarchy is a given, and profile details such as gender, race, age and class are not merely data, but decisive measures in the social stratification ladder.

The social media community we slide so easily into (and leave behind the moment we logout) may comprise an imaginary, but it is not fake. The platform is not a simulacrum of the social. Social media do not mask the real. Neither its software nor its interface are ironic, multi-layered or complex. In this sense, social media are no longer (or not yet) postmodern. The paradoxes at work here are not playful. The applications do not appear to us as absurd, let alone Dadaist. They are self-evident, functional, even slightly boring. What we find compelling is not the performativity of the interfaces themselves (which seems to be the feature of VR, now in its second hype cycle, 25 years after its first appearance). No, what attracts us is the social, the neverending flow.

Networks are not merely places of competition between rival social forces. This is a far too idealized point of view. If only. What particularly fails in this viewpoint is the notion of staging. Platforms are not stages; they bring together and synthesize (multimedia) data, yes, but what is missing here is the (curatorial) element of human labor. That is why there is no media in social media. The platforms operate because of their software—automated procedures, algorithms and filters—not through a large staff of editors and designers. Their lack of employees is their essence. It's what makes current debates about racism, antisemitism and Jihadism on social media so futile. Forced by politicians, social media platforms are now employing armies of editors ("cleaners") to do the all-too-human work of monitoring and moderating, filtering out supposedly ancient ideologies that have refused to disappear (more on platforms in Chapter 5).

Whereas gadgets such as smart phones and cameras have a (hyped-up and thus ultimately limited) fetish quality, the social network fails to register as having the same kind of status. Social media's power is due to its very banality. The network has become ecological, comparable to Sloterdijk's theory of the spheres. It surrounds us like air. It is a *Lebenswelt*, a (filter) bubble, an invisible dome comparable to the medieval worldview and the imagined Mars colonies. All ancient beliefs apply and have their legitimacy, from Plato's cave to Leibniz's closed monad. Pick a narrative; it applies to our social media reality. This also counts for the ideology take. Today's cosmology consists of layers of dating apps, soccer portals, software forums, videogames, and television sites like Netflix all woven together by search engines, news sites and social media. As in the case of air, proving the existence of this ubiquitous environment will be quite a task. But once the ideology reveals its ugly side, therapy works through the unconscious, paradoxes start to fall apart and the ideology unravels.

Going back to 2004, Wendy Chun was occupied with the issue of metaphors when taking software seriously as a new kind of social realism: "Software and ideology fit each other perfectly because both try to map the material effects of the immaterial and to posit the immaterial through visible cues. Through this process the immaterial emerges as a commodity, as something in its own right." The details seem less interesting to deal with: "Users know very well that their folders and desktops are not really folders and desktops, but they treat them as if they were— by referring to them as folders and as desktops. This logic is, according to Slavoj Žižek, crucial to ideology." It's worth noting that the Facebook category of Friends has become a similar metaphor. We can surely say the same of the Facebook newsfeed or running a YouTube channel.

So what will happen when the audience becomes too much to deal with? More important than deconstructing surface appearances, in Chun's words, is recognizing that "ideology persists in one's actions rather than in one's beliefs. The illusion of ideology exists not at the level of knowledge but rather at the level of doing." Here the rhetoric of interactivity obfuscates more than it reveals. Users negotiate with interfaces, computations and controls. But these surfaces hide the functionality beneath, meaning that they can never interact directly enough to understand. The Like economy behind our smart devices is a particularly relevant social media example. What will happen when we reveal that we have never believed in our own Likes? That we never really liked you in the first place?

Let's appraise the bots and the "Like economy"[12] for what they are: key features of platform capitalism that capture value behind the backs of its users. Social media are neither a matter of taste or lifestyle as in a consumer choice; they are our technological mode of the social. In the past century, we would never have considered writing letters or making a telephone call a matter of taste. They were cultural techniques, massive flows of symbolic exchange. Soon after its introduction, social media transformed from a hyped online service into essential infrastructure, underpinning social practices equivalent to writing letters, sending telegrams, and telephoning. It is precisely at this junction of becoming infrastructure that we (re)open the ideology file.

3

Distraction and its Discontents

"Never get high on your own supply." Ten Crack Commandments— "The Other as Distraction: Sartre on Mindfulness" Open University lecture—"She never felt like she belonged anywhere, except for when she was lying on her bed, pretending to be somewhere else." Rainbow Rowell—"This content is not suitable for all advertisers."— "In my head I do everything right." Lorde—"15 years ago, the internet was an escape from the real world. Now, the real world is an escape from the internet." Noah Smith—"Do Not Feed the Platforms" t-shirt—"How can you learn from mistakes if you don't remember them?" Westworld—"My words don't matter and I don't matter, but everyone should listen to me anyway." Pinterest—"Stop Liking, Start Licking" ice cream advertisement— #ThisIsWhatAnxietyFeelsLike—"Kick that habit, man." W. Burroughs.

Networks are not quite pleasure domes.[1] Discontent grows around forms and causes: from Russia's alleged interference in the 2016 US presidential elections to founding Facebook president Sean Parker admitting that the site purposely gives users a short trigger, outed as "addiction by design". Parker confessed: "It's a social-validation feedback loop... exactly the kind of thing that a hacker like myself would come up with, because you're exploiting a vulnerability in human psychology."[2] Next is Justin Rosenstein, inventor of the Facebook "Like" button, who compares Snapchat with heroin. Or Leah Pearlman, a member of the same team, admitting that she too has grown disaffected with the "Like" button and similar addictive feedback loops.[3] Or take Chamath Palihapitiya, another former Facebook executive, who claims that social media is tearing society apart and recommends that people "take a hard break."[4] In *Anti-Social Media*, Siva Vaidhyanathan writes that Facebook engages us like a bag of chips.

It offers frequent, low-level pleasures. It rarely engages our critical faculties with the sort of depth that demands conscious articulation of the experience. We might turn to Facebook in a moment of boredom and look up an hour later, wondering where that hour went and why we spent it on an experience so unremarkable yet not unpleasant.[5]

After reading such stories, who wouldn't feel betrayed? Cynical reason sets in as we realize the tricks being played on us. The screens are not what they seem. Soon after any behavioral targeting is exposed, our biases are confirmed; and as these effects start to wear out, marketing departments go on the hunt for the next forms of perception management. When will social media move fully off the stage of world history? Is it ever going to end? This leads to the question: what tangible effects does awareness of organized distraction have? Let's describe the ups and down of social media sensibility in detail. We know we're pulled away, yet continue to be interrupted—that's distraction 2.0.

A similar discontent is felt in my own net criticism filter bubble. What to do once you've been cornered from all sides and must come to terms with this mental submission? What is the role of critique and of alternatives when such a desperate situation becomes ubiquitous? Take the cryptocurrency critics who must have felt they lost out on the Bitcoin craze, getting stuck instead with a bunch of lousy Facebook friends. Depression is a general condition, whether realized or unrealized. The Internet—is that all there is? Discontent with the cultural matrix of the twenty-first century inevitably moves from the "technology" label to a political economy of society-at-large. Let's put our collective inability to change the internet architecture in the context of the broader "democracy fatigue" and the rise of populist authoritarianism, as discussed in the 2017 anthology *The Great Regression*.[6] But we also need to be aware that there is a dark side to this understandable gesture. Critical analyses often, unwillingly, end up in moral judgment. Shouldn't we instead ask the uneasy question of why so many were lured into the social media abyss in the first place? Is it perhaps because of the "Disorganization of the Will" Eva Illouz talked about in her study *Why Love Hurts*?[7] Many defend the usefulness of Facebook, WhatsApp and Instagram while simultaneously expressing mixed feelings about the moral policing of CEO Mark Zuckerberg, an ambiguity that masks a widely felt inability to make lifetime decisions. This is what Illouz describes as "cool ambivalence", a new architecture of choice in which rational and emotional considerations blur, causing a crisis of commitment in the choice of partners. Some of the saddest songs are about lovers leaving you just when you need them most. And it is precisely this that we see in the social media debate: I want to leave but I can't; there's too much going on but it's boring; it's useful yet disgusting. If we dare to admit it, our addictions are filled with an emptiness at the prospect of life unplugged from the stream. I want to delete it all, but not now.

Dopamine is the metaphor of our age. The neurotransmitter stands for the accelerated up-cycles in our mood, the euphoric high before the inevitable crash. The flux on social media varies from outbursts of expectation to long periods of numbness. Social mobility is marked by similar swings. Good and bad fortune stumble across each other. Life goes its way, until you suddenly find yourself in an extortion trap, your device hijacked by ransomware. We move from intense experiences of collective work satisfaction, if we are at all lucky, to long periods of job uncertainty, filled with boredom. Our interconnected life is a story of growth spurts, followed by long periods of stagnation in which staying connected no longer serves any purpose. Constant psychological boosts keep you hooked. As a result, we're dead inside. We feel defeated, over-whelmed, stressed, anxious, nervous, stupid, silly, useless.[8] Mood swings are programed—steadily up in the morning, followed by a parabolic tumble in the afternoon.

Let's call it social hoovering: we're sucked back in, motivated by suggestive improvements in conditions that never materialize. Social media architectures lock us in, legitimated by the network effect that everyone is on it—at least we assume they must be. The certainty, still held a decade ago, that users behave like swarms, freely moving together from one platform to the next, has been proven wrong. Departure seems persistently futile. We have to know the whereabouts of our ex, the event calendars and social conflicts of old or new tribes. One may unfriend, unsubscribe, log off or block individual harassers, but the tricks that get you back into the system ultimately prevail. Blocking and deleting is considered an act of love for oneself, a self-protection mechanism against becoming hooked. Yet the idea of leaving social media altogether is beyond our imagination.

Our unease with "the social" starts to hurt. Lately, life seems over-whelming. We go silent, yet return before long. The fact that there's no exit or escape leads to anxiety, burnout or depression. In his *Small Philosophy of Digital Abstinence*, Dutch writer Hans Schnitzler describes the liberating withdrawal symptoms his students at the Amsterdam Bildung Academy experience when they discover the magical experience of walking through the park without having to take Instagram snapshots.[9] At the same time, we hear a growing chagrin with such New Age school of life responses to digital overload. Internet critics voice outrage over the instrumental use of behavioral science aimed at manipulating users, only to realize that their concerns end up as digital detox recommendations in self-mastery courses. Nothing much happens after the Alcohol Anonymous style confession of your distraction. How do we escape the salience trap? Should one be satisfied with a 10% reduction of time spent on devices? How long does it take until the effect has worn off? Are you too longing for that calming feeling of being swaddled, longing to get rid of that restlessness? Well-meant self-help advice becomes part of the problem as it merely mirrors the avalanche of applications aimed to create "a better version of you."[10] Instead, we should find ways to politicize the situation. A critical platform approach should, first of all, shy away from any solution based on the addiction metaphor: the online billions are not sick and I'm not a patient either.[11] The problem is not our lack of willpower but our collective inability to enforce change.

We face a return of the high-low distinction in society with an offline elite that has delegated its online presence to their personal assistants, in contrast with the frantic 99% that can no longer survive without 24/7 access, struggling with long commutes, multiple jobs and social pressures, juggling complex sexual relationships, friends and relatives with noise on all channels.

Another regressive tendency is the televisual turn of the web experience due to the rise of online video inside all platforms, the remediation of classic TV channels on internet devices and the rise of services such as Netflix. A Reddit *Shower Thought* put it this way: "Surfing the web has become like watching TV back in the day, just flicking through a handful of websites looking for something new on."[12] Social media as the new TV is part of a long-term erosion of the once celebrated participatory culture, a move from interactivity to interpassivity.[13] This world is massive but empty. What's left by those who do comment are the visible traces of collective outrage. We read what the trolls have to say, and swipe away the verbal filth in anger.

One of the unintended consequences of social media usage is the growing reluctance to have direct verbal exchanges. In a blog-post titled "I hate telephones" James Fisher complains about the dysfunctionality of call centers and labels all "synchronous" telecommunication inefficient: "Asynchronous textual communication is how everyone communicates remotely now. It's here to stay."[14] According to Fisher, killing the telephone would mean killing a big market. This is part of a silent revolution. There's no rage against the telephone—the most effective way to sabotage the medium is simply not to take calls anymore. Teens don't take calls because it's seen as stressful. During a visit to a vocational media college in Amsterdam, I was told that the school had recently introduced a communication class for digital natives after firms had complained that interns were incapable of talking on the phone to clients. In line with Sherry Turkle's findings,[15] the course trains the students on how to conduct a conversation on the phone and in real life.

During a dialogue, be it on the phone or sitting next to each other in a café, we take the hermeneutics route and spread out the conversation. That's the art of interpretation, when we indulge in the exegesis of a situation, posting or episode. It's an expansive semiotic landscape where meaning is not tied to commitment. Instead, it's all about decision avoidance, probing into the world of the possible. We get lost in time while we ask, explain, interrupt and wonder, guessing the meaning of the hesitations and body gestures of our conversation partner. Such an expansive experience is the exact opposite of the compression technique, made visible in the condensed form of the meme. These visual messages compress complex issues into one image and add an ironic layer, explicitly aiming to propagate a message that can be grasped in a split second, before we swipe it away and quickly move on to the next posting. Memes beg to be liked and shared, making distraction visible as in the case of the Distracted Boyfriend meme.[16]

"Please approach me, astonish me." No matter how perfect the technology, smooth and fast exchanges remain the exception as we bump into the harsh reality of the Other. Whenever a text message is sent to someone, there is an expectation to receive one back. This wait, also known as texpectation, is the long and painful experience of anticipating a reply. The electronic ghost of the Other haunts us, until it finally appears on the screen.[17] "Every time my phone vibrates, I hope it's you." As Roland Barthes observes, "to make someone wait is the constant prerogative of all power." It is always me.

The other one never waits. Sometimes I want to play the part of the one who doesn't wait; I try to busy myself elsewhere, to arrive late; but I always lose at this game. Whatever I do, I find myself there, with nothing to do, punctual, even ahead of time. The lover's fatal identity is precisely this: I am the one who waits.[18]

In the dark days after the initial excitement, social media no longer fills the void. Throughout these loveless days one feels flat, like a failure, with little emotion. Some get angry easily, with social anxiety on the rise. When mood stabilizers no longer work, and you no longer get dressed during the day, you know you've been hoovered. Swiping fingers assist in moving the mind elsewhere. Checking the smartphone is the present way of daydreaming. Unaware of our brief absence, we enjoy the feeling of being remotely present. We remember what it's like to feel. While checking status updates we're wandering off in our mind, the movement is reversed and, without notice, the Other enters our world. Getting our phones out during any idle moment, for short bursts, the anxiety doesn't go away. Like daydreaming, social media visits can be described as "a short-term detachment from one's immediate surroundings during which a person's contact with reality is blurred."[19] The second part of this Wikipedia definition, however, doesn't fit. Do we pretend to be somewhere else when we quickly swipe through messages in the elevator? Momentary social media scans may be an escape from the present reality, but can we say that it is done to withdraw into a fantasy? Hardly. We glance through the updates and incoming messages for the same reasons as we daydream—to erase boredom.

Should we, with Sigmund Freud, look at social media use as an expression of repressed instincts? Or rather read social media as flows of digital signs coming from dispersed tribe members? Does the psyche need to reassemble close social ties, restoring a sense of kinship in an age of thinly spread-out networks? Social media revives the lost tribe. We reassemble those close to us on our devices. Can we describe the online version of the social as a "secondary revision",[20] a reprocessing of all the complex operations in our busy everyday lives? This would allow us to overcome what Nathan Jurgenson described as "digital dualism": the real and virtual are not separate spheres but a highly integrated, hybrid experience. Could we read the intense social media usage in cafes, on the street, in trains, in the kitchen and in bed, as an altered form of consciousness, this time fed by the outside world? We demand to be elsewhere. Against the widespread calls for more bodily or spiritual presence, such a way of looking at social media would instead reframe the mass obsession as an invasion of the elsewhere, a tele-presence in an invisible outside sphere we could also call techno-telepathy.

Admit the envy: others have rewarding experiences from which you are absent. That's the Fear of Missing Out, resulting in a constant desire for engagement with others and the world. This jealous feeling is the shadow side of the desire to be in the tribe, at the party, breast-to-breast. They dance and drink, while you're out there, on your own, in the cold. There is also another aspect: the online voyeurism, the detached form of peer-to-peer surveillance culture that carefully avoids direct interaction. Online we watch, and are watched. Overwhelmed by a false sense of familiarity with the Other, we're quickly bored and feel the urge to move on. While still aware of our historical duty to contribute, upload and comment, the reality is a different one. We've transgressed back to news outlets and professional influencers: only a few know how to turn attention to their advantage.

When applications are no longer new, they turn into a habit. This is the moment when geeks, activists and artists vanish from the scene and parents, psychologists, data analysts and marketing experts take their place. In *Updating to Remain the Same,* Wendy Chun argues, "media matter most when they seem not to matter at all, that is, when they have moved from the new to the habitual."[21] Chun describes habits as strange, contradictory things, both inflexible and creative. Habit enables stability in a fundamentally changeable universe. Its repetitive nature is not seen as something bad. "Habit, unlike instinct, is learned, cultivated: it is evidence of culture in the strongest of the world."[22] According to Chun, habit is such a timely approach as "neoliberalism emphasizes empowerment and volunteerism."[23] Paradoxically, its policy of privatization destroys the private sphere, resulting in internet users being turned inside out, framed as private subjects exposed to the public.

Call it what you want, "habitual media" capitalize on the wish for anti-experience, sharing information within one's own filter bubble (which Chun describes with the term "homophily"). Decoupled from its radically Other newness factor, social media upholds the desire for a manageable, contained difference—difference that has already been disarmed. This also plays out on the interpersonal level. In his *Anaesthetic Ideology* essay, Mark Greif notes a crisis in experience: "Experience becomes piercing, grating, intrusive. It is no longer a prize, though it is the goal everyone else seeks. It is a scourge. All you wish for is some means to reduce the feeling."[24] As friends become emotionally over-demanding, we grow overly detached, valuing our self-defense mechanisms as positive. Once we no longer care, and the melodrama is gone, we give it a glance, "Like" it and continue swiping onward. Social anxiety wears out, flattening out into a mood of indifference in which the world still glides, but with a quality of numbness. When the world is emptied of meaning, we're more than ready to delegate experiences to friends. No hard feelings. As distance grows, jealousy dissipates into the background.

Dutch technology critic Tijmen Schep created a website to further investigate the term "social cooling" that tries to capture the long-term effects of living inside a reputation economy. Cooling describes the simple observation that if you are being watched, you change your behavior. "People are starting to realize that their 'digital reputation' could limit their opportunities," Schep asserts.[25] This leads to a culture of conformity, risk aversion and social rigidity. Resistance against this logic will require decommissioning algorithms and criminalizing data gathering. Only if data analysis services are no longer available, will there be a chance of collectively "forgetting" these cultural techniques and their dreadful long-term consequences. His conclusion: "Data is not the new gold, it is the new oil, and it is damaging the social environment." A recent Data Prevention Manifesto argues along similar lines. It's not enough to protect privacy through regulation; both data production and capture need to be prevented in the first place.[26] For Schep, privacy means the right to be imperfect. We need to design for freedom, a freedom that actively undermines the technological pressures to lead a predictable life. If this does not occur, we may find ourselves living under a regime of social credit. Welcome to the *Minority Report* Society, one in which deviancy prevention has already been internalized in such a way that prediction is no longer required.

Remember *Her*? In this 2013 film, the main character, a male experiencing a mid-life crisis, falls in love with his female AI called Samantha. What's shocking is not the presumed computational brilliance of the female artificial character, or the lucidity of having phone sex with robots, but the introverted conformity that comes with the mass uptake of personal AI friendships. Once the attention of the masses has turned inward and has become routine, why bother with one's appearance? This is not quite the trend we see in social media culture. The film is both a moral warning of narcissistic solitude and a comforting soulful story about machines that assist us in the difficult passage from one relationship to the next. What's striking are the uniform, clumsy, geeky clothes everyone is wearing. Spike Jonze, the film director, says: "Have you ever worn high-waist pants? When we were doing wardrobe fittings, I tried them on, and I was like, 'Oh, these feel good!' They feel kinda like you're being hugged." Sleek, timeless 1940s fashion makes us feel familiar and comfortable. "When you add things that aren't of this era, you wind up noticing them and it becomes really distracting," the costume designer of the film admits. Everyone carries large clumsy bags. In *Her*'s retro-future scenario we've conformed to a uniform life and shied away from diversity. Similar to today's social media use, we can't say the subjects of *Her* are absent minded. The artificial interiority they inhabit, being structurally inattentive to outer things, shields off contact with the outside, much like the innocent *Hello Kitty* dresses that have been dominating the streets of metropolitan Asia for decades. Their positive commitment gives *Her* a dystopian taste.

In her book *Distributed Attention, a Media History of Distraction*, German media theorist Petra Löffler provides us with a relevant shift of perspective in this context.[27] Going back to the writings of Walter Benjamin and Siegfried Krakauer, she shows that distraction was once seen as a right that was claimed by the early labor movement. Repetitive factory work had to be compensated with entertainment. The demand for leisure time was supported by technologies such as the panorama, the world exhibition, the kaleidoscope, the stereoscope and the cinema—a metropolitan culture embodied in the figure of the gawker. Due to the rise of media technologies after World War II, this attitude slowly changed as the phase of disorientation set in.[28] As we've disconnected distraction from entertainment, we can no longer see the smart phone as a necessary toy for the reproduction of a labor force.[29] At what cost? Instead of policing digital daydreaming, we should bet on the horse called boredom. At some point, Silicon Valley will lose its war on attention and its ad-driven economy will inevitably start to slide. We're not there yet. Their strategies of behavioral fine-tuning and surprise still work.

Facebook fascinates. Löffler's move back in time could help free ourselves from the morals that surround the distraction discourse and instead ask what exactly is pulling us deeper and deeper into these networks. As Roland Barthes did with photography,[30] let's investigate what the "punctum" is in social media. How would you identify and then analyze the striking element that hurts and attracts you, that stands out, that rare detail your eye is searching for? It's the possibility of freedom and liberation from orchestrated stimulation, the unlikely information that will take us out of our routine. The irony here is that this relentless search results in a contradictory sense of repetition. What we desire is the next wave of disruptions—while simultaneously feeling unable to disrupt our own behavior. We're locked into a situation that makes it impossible to disrupt the disruptors.

As the discontent with the distraction discourse spreads, there's a growing revolt against the suggestion that it's all our own problem. Take Catherine Labiran, who no longer wants self-care to be seen as synonymous with pampering, complaining that she "grew tired of conversations about self-care being solely linked to some form of meditation."[31] According to Dutch media philosopher Miriam Rasch, with whom I have the privilege to work at the Institute of Network Cultures, digital detox therapy only fights the symptoms.

It overlooks the causes of perpetual distraction, loss of concentration and burn-outs. Going out into the woods without a phone to get relieved of stress will not help you in the long run. It's like the carrot in front of the donkey's nose: something that keeps you going, supposedly out of free will, while it's in fact a function of what Byung-Chul Han calls psycho-politics, the next step after Foucault's bio-politics. It means the psyche is in itself subjected to control mechanisms, which according to Han follow neoliberal rules. "The push towards self-discipline", of which digital detoxing is an example, is one of many strategies of the market to enter the psyche in order to increase efficiency, productivity, and profit.

According to Rasch, distraction is the first step in this process. "Once distraction has grown so disproportionately that we start to protest against it, detoxing and other disciplinary strategies are proposed as a second step, all the while helping corporations make more money." But Rasch is not willing to give up on the internet:

Apart from the negative "symptoms", it still offers a lot of benefits such as pleasure, friendship, courtship, knowledge and work. We need a new way of coping with distraction, one that befits the "post-digital" age, one that acknowledges that the internet is not going to go away— and we don't want it to, either. I demand a strategy that doesn't just turn away from the benefits and turn inwards into meditation and mindfulness but confronts the post-digital condition head-on, sucks it in, wallows around in it, and still thrives.

Would it be possible to politicize our own distraction, Rasch asks:

> We should stop being pursued by things distracting you. What in the world calls my attention? Listen to what's pleasing your car. I'd stress the "my" in my attention. Don't let yourself get hooked by anything that's fishing for attention. Become aware of attention: it's what media companies seek, and by seeking it, they destroy it. I don't care if it's online or offline—the two are hardly distinguishable—I care if I care, and I care about many things.

Media scholar Michael Dieter disagrees and warns that it's too easy to condemn digital detox retreats as just a neo-liberal ruse. Echoing Peter Sloterdijk's *You Have to Change your Life*, he claims that

> reactions to even temporary disconnection are often quite extreme. The retreat at least highlights a need for collective practices and changing the environment of use; I'm not sure we should trust our individual interests to fight distraction alone. Why not approach things with a more experimental mindset? We're not good at recognizing the potential impurity of such exercises. In this respect, indeed, the post-digital might be a useful concept. Pure detox is a risky endeavor, as medical experts claim: it can strengthen the impulses or habits that we aim to get rid of. Hybrid media experiences, diversified interdisciplinary forms of training and more-than-digital methods are some paths forward, along with a willingness to experience crisis as moments of clarity.[32]

The global elite is in two minds about the distraction epidemic, a confusion with profound implications for educational standards and pedagogical approaches. The rulers dream about totalitarian measures to overcome the current education crisis, measures that would somehow combine two distinctly different modes: fast-changing, distractive digital skills and reflexive deep learning methods. It is not in their interest to bring the hollow user to life. We're not just talking about doubts rationalized as ethical issues; the attention issue goes to the core of how the global economy is being shaped. On the one hand, one report after the other promises that considerable productivity gains will be made once there's no longer access to social media during work hours. On the other, a growing amount of businesses benefit precisely from the blurred boundaries between work and private life, from the precarious conditions of 24/7 availability that make permanent access a prerequisite and going offline a potentially dangerous affair. To put it in Stiegler's terms: the app that hooks us will also set us free.[33] Should the earlier "access for all" demand be updated to become the "right to disconnection"? Can we move beyond this dichotomy?[34] Existing social media lack hubris, style and enigma. It's their petty, sleazy, behind-our-back mentality that needs to be attacked. In order to overcome inevitable offline romanticism, we could ask: what's vital information for us,[35] how can we guarantee it reaches us through various filters, and to what extent do we accept built-in delays? Can vital information become air-gapped and get to us, even when we're no longer present on the networks?[36] How can we organize our social life in such a way? Whether offline or online, what counts is how, together, we might escape from a calculated life altogether. It was fun while it lasted, but now we're moving on.

4

Sad by Design

"Solitary tears are not wasted." René Char—"I dreamt about autocorrect last night." Darcie Wilder—"The personal is impersonal." Mark Fisher —Motivational speaker: "Swipe left and move on."—"I'm easy but too busy for you" t-shirt—"Why don't you just meet me in the middle? I am losing my mind just a little." Zedd, Maren Morris, Grey—"As the spirit wanes, the form appears." Charles Bukowski—"I don't care, I love it." Icona Pop—"Percent of riders on Shanghai subway staring at their phones: 100%." Kevin Kelly—"When you get ignored long enough you check peoples 'last seen' status to make sure they aren't dead." Addie Wagenknecht—"I don't feel like writing what I have just written, nor do I feel like erasing it." Kierkegaard—"The very purpose of our life is to seek happiness." Dalai Lama.

Try and dream, if you can, of a mourning app. The mobile has come dangerously close to our psychic bone, to the point where the two can no longer be separated. If only my phone could gently weep. McLuhan's "extensions of man" has imploded right into the exhausted self.[1] Social media and the psyche have fused, turning daily life into a "social reality" that—much like artificial and virtual reality—is overtaking our perception of the world and its inhabitants. Social reality is a corporate hybrid between handheld media and the psychic structure of the user. It's a distributed form of social ranking that can no longer be reduced to the interests of state and corporate platforms. As online subjects, we too are implicit, far too deeply involved. Social reality works in a peer-to-peer fashion. It's all about you and your profile. Likes and followers define your social status. But what happens when nothing can motivate you anymore, when all the self-optimization techniques fail and you begin to carefully avoid these forms of emotional analytics? Compared to others your ranking is low—and this makes you sad.

In *Ten Arguments For Deleting Your Social Media Accounts Right Now*, Jaron Lanier asks, "why do so many famous tweets end with the word 'sad'?"[2] He associates the word with a lack of real connection. "Why must people accept manipulation by a third party as the price of a connection?" According to Lanier, sadness appears in response to "unreasonable standards for beauty or social status or vulnerability to trolls." Google and Facebook know how to utilize negative emotions more readily, leading to the new system-wide goal: find personalized ways to make you feel bad. There is no single way to make everyone unhappy. Sadness will be tailored to you. Lanier noticed that certain online designs made him unhappy because social media placed him in a subordinate position. "It's structurally humiliating. Being addicted and manipulated makes me feel bad... There was a strange, unfamiliar hollow in me after a session. An insecurity, a feeling of not making the grade, a fear of rejection, out of nowhere."

Lanier discovered his inner troll, a troll produced by what he calls the asshole amplification technology: "I really don't like it when a crowd judges me casually, or when a stupid algorithm has power over me. I don't like it when a program counts whether I have more or fewer friends than other people." He refuses to be ranked and concludes: "The inability to carve out a space in which to invent oneself without constant judgment; *that* is what makes me unhappy." A similar reference we find in Adam Greenfield's *Radical Technologies* where he notices that "it seems strange to assert that anything as broad as a class of technologies might have an emotional tenor, but the internet of things does. That tenor is *sadness*... a melancholy that rolls off it in waves and sheets. The entire pretext on which it depends is a milieu of continuously shattered attention, of overloaded awareness, and of gaps between people just barely annealed with sensors, APIs and scripts." It is a life "salvaged by bullshit jobs, over-cranked schedules and long commutes, of intimacy stifled by exhaustion and the incapacity by exhaustion and the incapacity or unwillingness to be emotionally present."[3]

Of course sadness already existed before social media. And even when the smart phone is safely out of reach, you can still feel down and out. Let's step out of the determinist merry-go-round that all too quickly spins from capitalist alienation and disastrous states of mind to blaming Silicon Valley for your misery. Even technological sadness is a style, albeit a cold one. The sorrow, no matter how short, is real. This is what happens when we can no longer distinguish between telephone and society. If we can't freely change our profile and feel too weak to delete the app, we're condemned to feverishly check for updates during the brief in-between moments of our busy lives. In a split second, the real-time machine has teleported us out of our current situation and onto another playing field filled with mini reports we quickly have to investigate.

Omnipresent social media places a claim on our elapsed time, our fractured lives. We're all sad in our very own way.[4] As there are no lulls or quiet moments anymore, the result is fatigue, depletion and loss of energy. We're becoming obsessed with waiting. How long have you been forgotten by your loved ones? Time, meticulously measured on every app, tells us right to our face. Chronos hurts. Should I post something to attract attention and show I'm still here? Nobody likes me anymore. As the random messages keep relentlessly piling in, there's no way to halt them, to take a moment and think it all through.[5]

Delacroix once declared that every day which is not noted is like a day that does not exist. Diary writing used to fulfill that task. Elements of early blog culture tried to update the diary form for the online realm, but that moment has now passed. Unlike the blog entries of the Web 2.0 era, social media have surpassed the summary stage of the diary in a desperate attempt to keep up with real-time regime. Instagram Stories, for example, bring back the nostalgia of an unfolding chain of events— and then disappear at the end of the day, like a revenge act, a satire of ancient sentiments gone by. Storage will make the pain permanent. Better forget about it and move on.

It's easy to contrast the relentless swing between phone and life with the way anthropologists describe metamorphosis. Initiation and ritual are slow events that require time, instigated by periods of voluntary solitude. The perpetual now that defines the "smart" condition is anything but an endurance test. By browsing through updates, we're catching up with machine time—at least until we collapse under the weight of participation fatigue. Organic life cycles are short-circuited and accelerated up to a point where the personal life of billions has finally caught up with cybernetics. Time to go soft, *despacito*.

In the online context, sadness appears as a short moment of indecisiveness, a flash that opens up the possibility of a reflection. The frequently used "sad" label is a vehicle, a strange attractor to enter the liquid mess called social media. Sadness is a container. Each and every situation can potentially be qualified as sad. Through this mild form of suffering we enter the blues of being in the world. When something's sad, things around it become grey. You trust the machine because you feel you're in control of it. You want to go from zero to hero. But then your propped-up ego implodes and the failure of self-esteem becomes apparent again. The price of self-control in an age of instant gratification is high. We long to revolt against the restless zombie inside us, but we don't know how. Our psychic armor is thin and eroded from within, open to behavioral modifications. Sadness arises at the point when we're exhausted by the online world.[6] After yet another app session in which we failed to make a date, purchased a ticket and did a quick round of videos, the post-dopamine mood hits us hard. The sheer busyness and self-importance of the world makes you feel joyless. After a dive into the network, we're drained and feel socially awkward. The swiping finger is tired and we have to stop.

Sadness expresses the growing gap between the self-image of a perceived social status and the actual precarious reality. The temporary dip, described here under the code name "sadness", can best be understood as a mirror phenomenon of the self-promotion machine that constructs the links for us. The mental state is so pervasive, the merging of social media with the self so totalizing, that we see the sadness complex as a manifestation of an "anti-self" stage that we slip into and then walk away from.[7] The anti-climax called sadness travels with the smart phone; it's everywhere. It is considered sad when most of your friends are bots. The conservative judgment that many friends indicate a lack of character and gestalt[8] falls short here, as most are machine generated social relationships anyway. As buying followers has become more acceptable, social status no longer has to be built from the ground up through hard online labor.[9]

We should be careful to distinguish sadness from anomalies such as suicide, depression and burnout. Everything and everyone can be called sad, but not everyone is depressed.[10] Much like boredom, sadness is not a medical condition (though never say never because everything can be turned into one). No matter how brief and mild, sadness is the default mental state of the online billions. Its original intensity gets dissipated. It seeps out, becoming a general atmosphere, a chronic background condition. Occasionally—for a brief moment—we feel the loss. A seething rage emerges. After checking for the tenth time what someone said on Instagram, the pain of the social makes us feel miserable, and we put the phone away. Am I suffering from the phantom vibration syndrome? Wouldn't it be nice if we were offline? Why's life so tragic? He blocked me. At night, you read through the thread again. Do we need to quit again, to go cold turkey again? Others are supposed to move us, to arouse us, and yet we don't feel anything anymore. The heart is frozen. Once the excitement wears off, we seek distance, searching for mental detachment. The wish for "anti-experience" arises, as Mark Greif has described it. The reduction of feeling is an essential part of what he calls "the anaesthetic ideology". If experience is the "habit of creating isolated moments within raw occurrence in order to save and recount them,"[11] the desire to anaesthetize experience is a kind of immune response against "the stimulations of another modern novelty, the total aesthetic environment."[12]

Most of the time your eyes are glued to a screen, as if it's now or never. As Gloria Estefan wrote: "The sad truth is that opportunity doesn't knock twice." Then, you stand up and walk away from the intrusions. The fear of missing out backfires, the social battery is empty and you put the phone aside. This is the moment sadness arises. It's all been too much, the intake has been pulverized and you shut down for a moment, poisoning him with your unanswered messages. According to Greif, "the hallmark of the conversion to anti-experience is a lowered threshold for eventfulness." A Facebook event is the one you're interested in, but do not attend. We observe others around us, yet are no longer part of the conversation: "They are nature's creatures, in the full grace of modernity. The sad truth is that you still want to live in their world. It just somehow seems this world has changed to exile you."[13] You leave the online arena; you need to rest. This is an inverse movement from the constant quest for experience. That is, until we turn our heads away, grab the phone, swipe and text back. God only knows what I'd be without the app.

Los Angeles theorist and artist Audrey Wollen has declared sadness a feminist strategy, a form of political resistance "to be as goddamn miserable as we want."[14] In a text called *Sad Girl Theory*, she states, "our pain doesn't need to be discarded in the name of empowerment. It can be used as a material, a weight, a wedge, to jam that machinery and change those patterns." To Wollen, political protest is usually defined in masculine terms, "as something external and often violent, a demonstration in the streets, a riot, an occupation of space." Such a definition excludes "a whole history of girls who have used their sorrow and their self-destruction to disrupt systems of domination." Feminism doesn't need to advocate how awesome and fun being a girl is. The endless preaching of empowerment may as well be what Lauren Berlant calls a form of "cruel optimism". Sharing feelings online is not a form of narcissism. As Wollen insists: "Girls' sadness is not passive, self-involved or shallow; it is a gesture of liberation, it is articulate and informed, it is a way of reclaiming agency over our bodies, identities, and lives."

By reading sadness through a gender lens and contextualizing affect as a female response, Wollen turns sadness into a political weapon. And yet, in one sense, this weapon has already been defused. Today sadness has been compressed into code, turning it into a techno-sentiment. Audrey Wollen admits that social media ultimately abuses feelings with the aim of a positive quantifiable outcome. "Sadness has become quippy," she writes.

> I can tweet about how depressed I am instead of writing a sonnet in iambic pentameter. We spend a lot of time talking about how we want to kill ourselves over social media, but when was the last time all of your friends got together and cried? We still participate in upholding the idea of "happiness" as a goal or object that can be worked for, something you "earn" instead of just chilling with our misery.[15]

Sadness has neighboring feelings we can check out. There is the sense of worthlessness, blankness, joylessness, the fear of accelerating boredom, the feeling of nothingness, plain self-hatred while trying to get off drug dependency, those lapses of self-esteem, the laying low in the mornings, those moments of being overtaken by a sense of dread and alienation, up to your neck in crippling anxiety, there is the self-violence, panic attacks, and deep despondency before we cycle all the way back to reoccurring despair. We can go into the deep emotional territory of the Russian *toska*.[16] Or we can think of online sadness as part of that moment of cosmic loneliness Camus imagined after God created the earth. I wish that every chat were never ending. But what do you do when your inability to respond takes over? You're heartbroken and delete the session. After yet another stretch of compulsory engagement with those cruel Likes, silly comments, empty text messages, detached emails and vacuous selfies, you feel empty and indifferent. You hover for a moment, vaguely unsatisfied. You want to stay calm, yet start to lose your edge, disgusted by your own Facebook Memories. But what's this message that just came in? Strange. Did they respond?

Anxieties that go untreated build up to a breaking point. Yet unlike burnout, sadness is a continuous state of mind. Sadness pops up the second events start to fade away—and now you're down in the rabbit hole once more. The perpetual now can no longer be captured and leaves us isolated, a scattered set of online subjects. What happens when the soul is caught in the permanent present? Is this what Franco Berardi calls the "slow cancellation of the future"? By scrolling, swiping and flipping, we hungry ghosts try to fill the existential emptiness, frantically searching for a determining sign—and failing. When the phone hurts and you cry together, that's technological sadness. "I miss your voice. Call, don't text."[17]

SAD BY DESIGN OCCURRENCES

The hollow ache of sadness hurts. Yet it's also important to locate it empirically, to investigate its specific conditions. Far from being a natural response, such sadness is integrated into the design of interfaces and the architectures of apps. Today sadness has become technical, a shift that applies equally to users and producers. Let's first look at online video. Julia Alexander has documented the burnouts, panic attacks and other mental health issues of YouTube's top creators. Alexander reports, "constant changes to the platform's algorithm, unhealthy obsessions with remaining relevant in a rapidly growing field and social media pressures are making it almost impossible to continue creating at the pace both the platform and audience want." "This is all I've ever wanted. Why am I so unhappy?"[18] the 19-year-old YouTuber Elle Mills once cried out, echoing the earlier breakdown of Britney Spears in front of a television audience. Her life had changed so fast, that it resulted in a breakdown in front of the camera. While daily television shows have large crews with editors and studio spaces, vloggers often broadcast out of their own apartments, producing clips on their own or with a small crew. And whereas TV hosts receive famous guests and deal with societal issues, YouTube celebs are more likely to report on their own ups and downs. Millennials, as one recently explained to me, have grown up talking more openly about their state of mind. As work/life distinctions disappear, subjectivity becomes their core content. Confessions and opinions are externalized instantly. Individuation is no longer confined to the diary or small group of friends, but is shared out there, exposed for all to see.

"When the careers of so many video personalities involve exposing their personal lives, striking a work/life balance is next to impossible," Alexander notes. Keeping up the vlogs is hardly a voluntary choice. If you take a break, even for a day, you immediately drop in the algorithm rank that favors frequency and engagement. We're dealing here with pre-programed mental breakdowns, exhaustion directly brought on by software settings, collapse coded in by developers under the supervision of senior engineers. "No one is telling YouTubers to chill out," Alexander concludes. "It's the opposite. People constantly ask for more, and there's only so much that one person can offer."

A next case would be Snapstreaks, the best friends fire emoji next to a friend's name indicating that "you and that special person in your life have snapped one another within 24 hours for at least two days in a row."[19] Streaks are considered a proof of friendship or commitment to someone. So it's heartbreaking when you lose a streak you've put months of work into. The feature all but destroys the accumulated social capital when users are offline for a few days. The Snap regime forces teenagers, the largest Snapchat user group, to use the app every single day, making an offline break virtually impossible.[20] While relationships amongst teens are pretty much always in flux, with friendships being on the edge and always questioned, Snap-induced feelings sync with the rapidly changing teenage body, making puberty even more intense.

Evidence that sadness today is designed is overwhelming. Let's take the social reality of the WhatsApp billions seriously; these are not some small-town plodders. The grey and blue tick marks alongside each message in the app may seem a trivial detail, but let's not ignore the mass anxiety it's causing. Forget being ignored. Forget pretending you didn't read a friend's text. Some thought that this feature already existed, but in fact two grey tick marks signify only that a message was sent and received—not read. The user thinks: "My message was delivered. I read in airplane mode." A site explains: "Once this mode has been enabled, the user can then open the app and read the message without alerting the sender's attention to their action by triggering the blue ticks."[21] Your blue tick marks haunt me in my sleepless nights. Those blue ticks.[22]

In response to rising anxiety levels, WhatsApp provided a list of reasons why someone may not have yet received your message. Their phone might be off; they could be sleeping, especially if they live in a different time zone; they might be experiencing network connection issues; they might have seen the notification on their screen but did not launch the app (especially common if the recipient uses an iPhone); and most importantly, they might have blocked you—just in case you were wondering what happened. There may be a temporary inability to communicate. You keep opening the app in the hope of finding something good, even though you know you are going to find nothing. You keep guessing and go mad. "You are craving for some appreciation, love, respect, attention which you are not getting in the real world, hence you are having an expectation from a virtual world that somebody may admire/like/respect you, due to these expectations you get anxious and get worked up as those things rarely or never happen!"[23] This is online despair, the worst trip ever: "It's easier to deal with not knowing why someone isn't replying, than to deal with repeatedly questioning why someone had read your message but refused to reply."[24]

Even if you know what the double tick syndrome is about, it still incites jealousy, anxiety and suspicion. It may be possible that ignorance is bliss, that by intentionally *not* knowing whether the person has seen or received the message, your relationship will improve. The bare-all nature of social media causes rifts between lovers who would rather not have this information. But in the information age, this does not bode well with the social pressure to participate in social networks. The WhatsApp color feature might also expose the fatal flaws in an emerging relationship—for some, this may be a way to dodge a bullet. One response is to change the settings and disable the color function so that no more blue ticks show up after you read a message, shunting all communication into the ambiguous zone of the grey tick. This design is for dummies. You may not understand a thing about the technicalities of wi-fi or algorithms, but it's damn easy to grasp the relational stakes of the double check syndrome. "You obviously read it, so why didn't you respond?"

The last case discussed here centers around dating apps like Tinder. These are described as time killing machines—the reality game that overcomes boredom, or alternatively as social e-commerce—shopping my soul around. After many hours of swiping, suddenly there's a rush of dopamine when someone likes you back. The goal of the game is to have your egos boosted. If you swipe right and you match with a little celebration on the screen, sometimes that's all that is needed. "We want to scoop up all our options immediately and then decide what we actually *really* want later."[25] On the other hand, crippling social anxiety is when you match with somebody you are interested in, but you can't bring yourself to send a message or respond to theirs "because oh god all I could think of was stupid responses or openers and she'll think I'm an idiot and I am an idiot and…"

Sherlyn from Singapore talks about one of her experiences on that lonely sea called OKCupid:

> I am not entirely sure why I venture in and out of this site. I always feel at once gutted and hopeful. I have chatted with many, but never have actually met anyone. I am highly anxious of translating anything to the real world. Where is this anxiety coming from? Is it the rejection I am worried about, or in fact falling into the trap of it?

In another instance, Sherlyn started chatting with a person who claimed to be a documentary filmmaker for humanitarian organizations.

> It appealed to me. We started mailing, and I sent him a link to my profile on academia, just as a way to put myself out there and asked more specific and pointed questions about his work. He responded: "This sounds more like a job interview than meeting on OKC." I got the message and responded with: "My work is what defines my politics, passion, and poetic, and it is perhaps the only way I can define my being. I can sense that you are expecting something else, considering where we met, thus I suggest you move on. Thanks." His response was rather prompt: "I don't have time for politics, go waste someone else's time, you political whore and slut."[26]

NO MELANCHOLY FOR YOU

Let's compare *fleeting* sadness in its technical form with the ancient state of melancholy. The melancholic personality seems to suffer from a disease. Unable to act, she withdraws from the world, contemplating death and other transient phenomena. While some read this condition as depression and boredom, others reframe this lazy passivity as a creative strategy, waiting for inspiration to strike. Instead of a fascinating *dérive* into the vast arsenal of literary sources, I propose here a digital hermeneutics that short-circuits philology with the eternal presence of the digital that surrounds us.

Take Susan Sontag's musings on Walter Benjamin as a man beset by a profound sadness, *un triste*.[27] As Benjamin wrote: "I came into the world under the sign of Saturn—the star of the slowest revolution, the planet of detours and delays...." Compare this deep, lingering melancholy with the snark we receive from others in response to a selfie with a friend, and the way it troubles us to no end.[28] How do today's "children of Saturn" (that planet of detours) deal with the unbearable lightness of the social that turned reflection into a rare state of exception? It's not quite *un bonheur d'être triste*. Nor does it quite match the classic boredom German style—the feeling you hate everything.

Melancholy, often described as sadness without a cause, has strong existential connotations. While paying tribute to Kierkegaard, who liberated melancholia once and for all of its medical stigma, describing it as the deepest foundation of the human in a Godless society, the problem here is not a vertical one of going deeper, but a horizontal one. The democratization of sadness happens through its thin spread across our plateau—homeopathic doses flatly distributed via technical means. Ever since antiquity, melancholia has been described as either something natural, rooted in the human condition, or as a chronic disease, brought on by heavy meals and dark red wines. In *Problemata XXX.1*, Aristotle brings the constitution of the fluids, the dry and the wet, in relation with hot and cold temperatures of the body.[29] The proposal here is to add a next layer: the technical temperament. For centuries, melancholia has been conceived as a gloomy state of mind. While ancient descriptions explain that the gloom stems from a particular mix of black and yellow bile, blood and phlegm, we could update this diagnosis to include blue bale, the color of our saturnine apps.[30]

And yet if fluids keep on flowing, they may no longer be the best way to analyze our sociotechnical condition. The metric to measure today's symptoms would be time—or attention, as it is called in the industry. While for the archaic melancholic, the past never passes, techno-sadness is caught in the perpetual now. Forward focused, we bet on acceleration and never mourn a lost object. The primary identification is there, in our hand. Everything is evident, on the screen, right in your face. While confronted with the rich historical sources that dealt with melancholia, the contrast with our present condition becomes immediately apparent. Whereas melancholy in the past was defined by separation from others, reduced contacts and reflection on oneself, today's *tristesse* plays itself out amidst busy social (media) interactions. In Sherry Turkle's phrase, we are alone together, as part of the crowd—a form of loneliness that is particularly cruel, frantic and tiring.

What we see today are systems that constantly disrupt the timeless aspect of melancholy.[31] There's no time for contemplation, or *Weltschmerz*. Social reality does not allow us to retreat.[32] Even in our deepest state of solitude we're surrounded by (online) others that babble on and on, demanding our attention. But distraction does not just take us away from the world—this is the old, if still prevalent way of framing the fatal attraction of smart phones. No, distraction does not pull us away, but instead draws us back into the social. Social reality is the magic realm where we belong. That's where the tribes gather, and that's the place to be—on top of the world. Social relations in real life have lost their supremacy. The idea of going back to the village mentality of the place formerly known as real life is daunting indeed.

SO SAD TODAY

Social media anxiety has found its literary expressions, even if these take decidedly different forms than the despair on display in Franz Kafka's letters to Felice Bauer. The willingness to publicly perform your own mental health is now a viable strategy in our attention economy. Anyone who can bundle up the dreary processes of living into an entertaining package develops at least the prospect of monetization and celebrity. Take the US writer Melissa Broder, who joined Twitter in 2012 with her *So Sad Today* account after she moved from New York to Los Angeles. Her "twitterature" benefitted from her previous literary activities as a poet. Broder has mastered the art of the aphorism like few others, compressing feelings and anxieties into bite-sized tweets.

Broder writes about issues such as low self-esteem, botox and addiction in an emotional manner. She is the contemporary expert in matters of apathy, sorrow and uselessness. During one afternoon she can feel compulsive about cheesecakes, show her true self as an online exhibitionist, be lonely out in public, babble and then cry, go on about her short attention span, hate everything and desire "to fuck up life". Internet obsession is her self-obsession. In between taking care of her sick husband and the obligatory meeting with Santa Monica socialites, there are always more "insatiable spiritual holes" to be filled. The more we intensify events, the sadder we are once they're over. The moment we leave, the urge for the next experiential high arises. Fashion magazine *Elle* has called Broder "Twitter's reigning queen of angst, insecurity, sexual obsession and existential terror."[33] Others have labeled her as yet another worker in the "first-person industrial complex."[34] I would call her the ideal *Internetgesamtsubjekt*.

After having a suicide vision on a Venice Beach sidewalk, Lucy, the main character in Melissa Broder's 2018 novel *The Pisces*, suddenly became afraid. "I took out my phone and pressed the buttons to get a car to take me home. This was just what people did now. We went from emotion to phone. This was how you didn't die in the twenty-first century." As phone and life can no longer be separated, neither can we distinguish between real and virtual, fact or fiction, data or poetry. In Broder's universe it's all part of one large delirium, an inexorable spiral downwards. "What I have sought in love is a reprieve from the itch of consciousness." She sums up her "lifetime of fictional love stories" through the veil of her insecurities. In her book of essays *So Sad Today*, we find Twitter or SMS-length messages that all end with ": a love story". "Sorry I fell asleep while you were going down on me: a love story." "I've been on your FB page for five hours today: a love story." "I don't even masturbate to you anymore because it's too sad: a love story." "I don't want to get off the internet or consider anyone else's needs: a love story." "When I send nudes, I like to receive a full dissertation on their greatness: a love story." "We're going to spend the rest of our lives in my head: a love story." "No teeth on the clit, thanks: a love story." "Tell me if I'm texting too much: a love story."[35]

Another episode in *So Sad Today* deals with a not-so-imaginary internet love affair. It started off with

silly messages and praise for my writing and a picture drawn in my favorite candy. (..) He poked and messaged and liked my every Internet itch. One afternoon they started a sexting game, which takes up six pages of seductive, explicit language: "Him: I want to feel your moans on my cock. Me: I want you to tease my belly, pussy and thighs until I am begging.

The sexting continued for a year until they met in a Manhattan hotel. They met a few times, had sex in all colors of the rainbow, then came back to sexting, but that too was now ruined by reflection—spiraling down into more sadness. They realized they could not have a normal relationship and broke up. "i have decided to give monogamy a try. This means the end for you and me in a sexual/textual way. i am deeply sad as i write this. we did so good. good love. another lifetime?)" After months of agony, she starts to write up the story. "What I maybe miss most is being able to lapse into space land and fantasize about the sex with him.

(..) I want to say: was I real to you? (..)" We got to be magic together. But is magic even real? She ends: "Online dating is sad. Attending holidays and weddings alone is sad. Marriage, too, is sad but love, lust, infatuation—for a few moments I was not sad."

Her tweets cover the spectrum from female sensibility to social anxiety: she despises modern life ("waking up today was a disappointment", "staying alive is a lot of fucking pressure"), hates herself ("i wouldn't fuck me"), is self-destructive ("a positive feeling can fuck you up forever", "i don't want to do what's good for me"), never pretends that life is better than it is ("i'm not moisturized, hydrated or full of self-love"), makes demands ("i don't think we get the dick we think we deserve", "don't tell me about the science of the brain just tell me how to feel better"). So Sad Today registers the widely felt numbness ("can't decide if i'm alive", "my drug of choice is low self-esteem"), is addicted to instantaneous changes ("fell in love with 8 people in 10 minutes"), lives the inevitable ("horoscope: you shouldn't text him but you will"), feels empty ("i've been awake 5 minutes and it's already too much") and judges others ("your positivity feels like a lie"), has suicidal tendencies ("i want to donate all my blood"), radicalizes human relationships ("being just friends is a nightmare"), is excellent in summarizing her ongoing short affairs ("loving you was an illness"), presenting her followers with a neverending stream of hypermodern dilemma's ("should i eat, nap or masturbate: the musical").[36]

Is Broder's sadness merely a literary effect that gives synthetic love a human touch? Broder's polyamorous relationship status is neither desperate nor liberating. There's a brutal honesty in the way she describes her multiple sexual relationships that reminds us of Michel Houellebecq. Is Broder's sadness merely a literary effect that gives synthetic love a human touch? We can contrast the Broder persona with the femme fatale in Amos Kollek's 1997 film Sue, a tragic New York tale of a déclassé secretary who's losing her job and apartment.[37] The medical metaphor of sex addiction in the movie here stands for economic decline. Two decades later there's not a trace of victimhood or poverty in Melissa Broder's work. The polyamorous lifestyle is already an integral part of the precarious condition. Instead of empathy, the cold despair invites us to see the larger picture of a society in permanent anxiety. If anything, Broder embodies Slavoj Žižek's courage of hopelessness: "Forget the light at the end of the tunnel—it's actually the headlight of a train about to hit us."[38]

MOURNING THE LOSS OF COMMUNICATION

The purpose of sadness design is, as Paul B. Preciado calls it, "the production of frustrating satisfaction."[39] Should we have an opinion about internet-induced sadness? How can we address this topic without looking down on the online billions, without resorting to fast-food comparisons or patronizingly viewing the public as fragile beings that need to be liberated and taken care of.[40] I am with Italian design theorist Silvio Lorusso who writes:

If design becomes just an expression of bureaucreativity hidden by an exhausting online and away-from-keyboard emotional labor, the refusal of work, of its bodily and cognitive dimension, should go hand in hand with the refusal of mandatory enthusiasm, of the positive disposition that such work requires. This is why my call for sadness is actually a plea for an emotional counterculture, a collective reaction against the occultation of material circumstances by means of artificial self-motivation. Fellow imposters, stop smiling and coalesce.[41]

Before we call, yet again, to overcome Western melancholy, it's important to study and deconstruct its mechanisms. In a design context, our aim would be to highlight "the process in which a designer focuses on the consequences of the current situation instead of dealing with the causes of a particular problem."[42]

We overcome sadness not through happiness, but rather, as Andrew Culp insisted, through a hatred of this world. Sadness occurs in situations where the stagnant "becoming" has turned into a blatant lie. We suffer, and there's no form of absurdism that can offer an escape. Public access to a twenty-first-century version of Dadaism has been blocked. The absence of surrealism hurts. What could our social fantasies look like? Are legal constructs such as creative commons and cooperatives all we can come up with? It seems we're trapped in smoothness, skimming a surface littered with impressions and notifications. The collective imaginary is on hold. What's worse, this banality itself is seamless, offering no indicators of its dangers and distortions. As a result, we've become subdued. Has the possibility of myth become technologically impossible? Instead of creatively externalizing our inner shipwrecks, we project our need for strangeness on humanized robots. The digital is neither new nor old, but—to use Culp's phrase—it will become cataclysmic when smooth services fall apart into tragic ruins. Faced with the limited possibilities of the individual domain, we cannot positively identify with the tragic manifestation of the collective being called social media. We can neither return to mysticism nor to positivism. The naïve act of communication is lost—and this is why we cry.[43]

5

Media Network Platform: Three Architectures

If sadness is individual, it is also architectural. The conditions that produce sadness are not constrained to the single self, but operate on a far broader level, surrounding and shaping society. Along with the personal experience of sadness, then, we also need to understand how it becomes operational through a wider set of infrastructures and environments. How does sadness scale?

For Michel Foucault, the hospital, asylum and prison symbolized disciplinary society. Today's institutions of self-containment are no doubt the social media platforms. In one sense, nothing has changed—these platforms are embedded with similar pedagogical intentions as the nineteenth century institutions. In another sense, everything has changed—these sociotechnical architectures have replaced institutions, challenging conventional forms of self-mastery and control. The key question in this chapter is how to unravel this architecture, how to take apart social media so that, in Foucault's words, the "obscured political violence within them would be unmasked."[1] My aim here is not to uncover the weak disciplinary form of social media. After all, distraction is not an escape from discipline. So before we rush into cookie-cutter critique—privacy concerns, monopolistic ownership and state surveillance measures—I would like to investigate the term platform itself. What's so comforting about being on the platform? To explore its allure, I'll compare "platform" with two earlier and broader terms significant in my own biography: media and network.

In both daily and scholarly language, the terms of media, network and platform have become interchangeable. Whereas television, newspapers and smart phones are easily identifiable as distinct material carriers, social media blur all boundaries into one fuzzy online experience. To put it in today's lingo: we share media on platforms through networks.[2] Each of the three concepts has been a *grande idée* that burst into the sphere of everyday language in a particular era. While the 1980s were the golden era of media, and networks dominated the nineties and the noughties, we're now well into the platform age. Ever since Trump, we're aware of just how easy it is for statements on platforms to become headline news. Everything slides together. Is this what the media business once predicted as convergence? What do we gain by insisting on the separation between such distinct forms of expression?

Technology is developing rapidly, but the academic classification of disciplines remains conservative and deeply rooted in past centuries. What's continental European media theory all about? And how does this subgenre of the humanities relate to Anglo-American cultural studies? Why is there no link from these half-baked disciplines to those who study media and communications? And how does the uncertain future of the dead star humanities fit into all of this? Whereas media studies seems to consolidate its position, no one seriously took on the challenge to establish Internet studies, let alone platform studies—despite over-whelming global statistics that illustrate a majority of mankind (55% in June 2018) are online—hooked on platforms.[3]

In fact, contemporary platforms claim neither to be media nor networks, stubbornly maintaining their engineering fantasy of merely operating as a technical provider. With such fast-moving targets and quickly changing labels, should we make a case for platform or app studies, knowing the tragic disappearance of new media as inter or trans-disciplinary research?

Despite the academic mess, it's not that difficult to chronicle how we went from media via networks to platforms. Just follow the different stages of the Internet's development. After its military and academic origins and its transformation into public infrastructure, the turn came in 1997. Suddenly the new media period—in which we, as activists, artists, designers and community organizers believed we could play a role—came to an abrupt end and the venture capital monoculture took over. "No rhizome for you." Obey the supermarket design. Succumb to the takeover. The "short summer of net criticism"[4] was followed by the dotcom hype of e-commerce, until its crash in 2001 (finished off by 9/11). In the wake of this bubble bursting, Web 2.0 was a low-key period of recovery. Blogs, RSS feeds and user-generated content took over the collective imaginary and Google began its success story. In this era, the knowledge of networks acquired in the previous periods was trans-formed into code—and into profit for the few. The fourth internet phase, in which we're still stuck, began after the 2008 global financial crisis and is defined by the rise of the extractivist model, social media platforms that subordinate networks as mere tools for hypergrowth.

In this chapter, I first tackle the question of media before turning to the uncertain status of the network and concluding with the current state of the platform. A chronology, like the one given above, certainly provides a powerful narrative that coincides with my own biography. Nonetheless, I prefer to explore these three terms as interwoven plateaus—infrastructural layers that form the larger framework of "the stack".

ETERNAL RETURN OF THE MEDIA QUESTION

My biographical point of departure to address the Media Question is not alternative media or Marshall McLuhan but German media theory. When I decided in 1987, the year I bought my first PC, to become a media theorist—without having much of an idea what this would imply outside of academia—Klaus Theweleit (*Male Fantasies*) and Friedrich Kittler (*Film-Gramophone-Typewriter*) were my role models. They were storytellers in the psychoanalytic tradition that investigated the traumatic roots of media in the Second World War. For these thinkers, media could not be separated from the military. Cybernetics was born from command and control. This rear window approach was radically different from IT's neverending cycles of hype and their obsession with the future. Darker and laced with violence, the historical materialism they presented was anything but nostalgic.

In 2004 Günter Helmes and Werner Köster published the *Texte zur Medientheorie* anthology in the historical yellow Reclam series. The 352-page booklet neatly encapsulates the discourse dominant in that part of Europe at the time. One point of difference is that, in contrast to many Anglo-Saxon readers with a post-colonial or feminist cultural studies angle, their approach cannot exactly be called politically correct. In fact, one could quite easily describe their scope as straight out conservative. Another point of difference is their emphasis on the materiality of the media. For example, the editors chose to include a fragment from Friedrich Albert Lange, who wrote his history of materialism in 1866. For Helmes and Köster, while media messages may represent one meaning or another, the actual content is irrelevant for media theory. This distinction has been revived in contemporary media activism: whereas some criticize certain media for their messages, others work on taking apart the media apparatus as such.

The editors are not exactly self-aware of their German bubble. The specificity of German media theory is never acknowledged; its particular problems and methods never contrasted with the Canadian school or the British-US-Australian cultural studies approach. Arguably media theory itself is a continental European humanities affair. As Florian Cramer has explained on numerous occasions, the metaphysics on display here arose from a specific crisis when German philosophy and humanities became aware of its own materiality (and futility).[5] On the plus side, we could praise the editors for their idiosyncratic weirdness. Media theory aimed at cultural superiority, philosophical supremacy over a strategic field, without having to compromise with mundane creative industries practitioners. The result was that German cultural and academic elites were given the (financial and conceptual) freedom to contemplate media in a unique and untimely manner.

This Reclam anthology doesn't offer complete texts; it is a collection of one- to three-page fragments, all 76 written by males. The collection starts off with the Biblical verses in Exodus regarding the prohibition of images. Flipping through Plato, Cicero and Augustine it becomes clear what the essence of media is all about: policing of the senses. For these thinkers, reading destroys your ability to memorize, to have meaningful dialogues, and so on. Then we fast-forward to the eighteenth century, a highly imaginative and innovative time period with a corresponding explosion of entries. The twentieth century, in comparison, offers fewer surprises. Starting off with early film theory, entries take in Weimar classics such as Brecht and Benjamin, move through to post-war technology debates, television, hypertextuality, media *an sich* and virtual reality, and finally close with the Bill Joy/Ray Kurzweil debate on AI and robotics.

The selected fragments deal with the art of fine-tuning the senses. A popular topic at the turn of the millennium, this was known at the time as multimedia—a shadow of Richard Wagner's *Gesamtkunstwerk*. Yet the German strategy of stretching out media to its maximum has not led to a general acceptance of media theory as an organizational structure to replace philosophy or religion as centralized meaning providers. At best, it provided a limited group of scholars with valuable insights into the history of culture and communication. What we're left with is an ongoing dominance of partial media with their peculiar distribution channels and particular industry titans. Think of film, television, radio, poetry, photography, newspapers, theatre, and book publishing. One day the Internet might also be added to this list (perhaps only when the planetary project of the "medium to end all media" completely fails, a totalizing fantasy collapsing like the Roman Empire). This is not another requiem for the media. It is far too late for such a gesture. Media theories may come and go, yet media are here to stay. These days, phenomena do not just disappear; they become part of the landscape, somehow surviving every new wave that promises to wash them into the tide of historical irrelevance. In the case of the media, we should not fall victim to the widely predicted relativism and indifference. From daily newspapers to local radio and national television, including websites, there is still a collective belief in the importance of the media. Niche, retro formats are just as real as the technological innovation that is supposed to render them obsolete or overwhelm them with abundance. Repackaging of content is now going in all directions. Politically too, media remain important, even if few comprehend their power. With the notable exceptions of Trump on Twitter or the Italy Five Star movement that emerged from the Web 2.0 era, political and intellectual elites remain clueless about the present possibilities. In the absence of a cultural avant-garde that is tech savvy, the political vision of the citizen-as-user remains unrealized and inadequate.

NETWORKS AS SECONDARY ORDER SYSTEMS

Let's face it: networks did not take over the world. Their auto-poetic dynamics, aimed at empowering the individual in societies where fixed social relationships were declining, was touching, but in the end overrated. As visual diagrams or architectural constructs, networks are convincing. As a sustainable economic or institutional framework, networks deliberately do not deliver. The inward-looking, feedback-driven nature of the network is both its strength and weakness. Before a network theory was able to develop and spread, it was sidelined by rhizomatic postmodern thinking, a discourse that walked away from the hard question of how networks were going to supersede the static formalism of twentieth century industrial relations. Was the historical collision of networks and postmodernism simply a coincidence, or a straight up mistake? Chalk this up as another question that will go unanswered.

Fast-forward to today, and what remains are two network legacies. The first is Manuel Castells' sociology of flows, a concept that became highly influential after his *Network Society* trilogy appeared at the height of dotcom mania. Castells' network society approach can be positioned in between social movement studies, internet research and urban studies. The second legacy is the network science school of Albert-László Barabási, Duncan J. Watts and others. Dating from the same late 1990s period as Castells, this approach is rooted in mathematics and computer science.[6] And yet neither legacy has built up its own research schools. Instead, they maintain a weak, almost invisible presence inside existing disciplines. This can be said of internet research in general. Paradoxically, both affirmative and critical network theory excel at undermining their own future. Practice what you preach.

Regardless of the poverty in academia, networks are now the major tool for any political and social work. Indeed, more than a tool, networks constitute their own *Umfeld* or environment, a sphere of activity we are often barely aware of. Networks define our horizon, and they are pretty wide these days. Networks surface when formal hierarchies are questioned or collapse altogether. They are hand-made products of human labor. One does not build them overnight. Unlike "friends", they cannot be purchased. Once the work is done, the network can, of course, be mapped, simulated and captured. But the origins of the network remains a mysterious a-priori. No matter how much computer power we have or how much visualization software we possess, diagramming this network ultimately generates a set of dead entities. Fixed and fossilized, it will always fail to capture the lively play of social impulses. At best, it may indicate that somewhere social dynamics took place. As Zeynep Tufeksi insists, "network internalities do not derive merely from the existence of a network but from the constant work of negotiation and interaction required to maintain the networks as functioning and durable social and political structures. Building such networks is costly."[7] She reminds us that "technology can help movements coordinate and organize, but if corresponding network internalities are neglected, technology can lead to movements that scale up while missing essential pillars of support."

So far the anthology that best summarizes the depth and scope of the network concept is *Networks*, edited by Lars Bang Larsen and published in 2014 by Whitechapel Gallery together with MIT Press. The starting point here is not a cultural history of the network as a myth or symbol, the collection of connections as described by someone like Sebastian Gießmann.[8] Instead, it starts in the fuzzy pre-dotcom period, an era rich with speculation from visionaries and artists. Here Sadie Plant wonders about "weaving the web", Joseph Beuys sends honey "flowing in all directions" in an exhibition for Documenta 6, and Felix Guattari muses on the connectedness of his three ecologies. Yet after the euphoria of the network "as a mode of being", the anthology quickly descends into the sections "exchange is the oxygen of capital" and "corruption, intrigue and covert solidarity." There's not much left of the initial excitement of being connected. Inside the Temporary Autonomous Zones there was drama and dirtiness. But even in those days, the nutty multitudes somehow sensed that the ecstasy of the "telematic embrace" was all too fleeting.

THE NEW COMMUNICATION DEFAULT: PLATFORMS

As we speak, platforms are superseding both media and networks. Theorization of platforms is still in its infancy. If we want to write a genealogy of the platform, we could start off with an etymology of the Dutch *platte vorm*, the "flat form" as a level surface that operates as a gigantic equalizer of different forces and streams.[9] Rather than the lower surface of the polder, it's important to note that the plateau is always situated on a higher level, following the ancient military strategy of positioning fortifications around churches, palaces and castles on hills in order to detect enemy movement. In the same way, platforms today harness informational differences and cannot easily be invaded. Platforms rule the surrounding territory, much as the Vauban fortresses once did. In principle they can be hacked and flooded, but in practice these disturbances are temporary (though the proviso here is that fully fledged cyber-warfare has not yet occurred). Alongside the fortification, platforms can also be compared with the city square, where markets were historically held. However, this metaphor should be stripped of any of the "natural" notions like supply and demand or the price equilibrium. Far from being natural, these environments are highly artificial—their economies invisibly manipulated by operators and "first movers" through algorithms and code.

Today, every ambitious person wants to starts a platform. Their worldwide success has inflated the term beyond all measure. Platforms are now seen as the architecture to emulate. But while this sounds like a worthy pursuit, we need to take into account that the platform as we've known it thus far is not just a successful website. To have a heavy social media presence is one thing; to build a platform is something of an entirely different order. Witness the daunting set of dependencies needed for a platform to emerge. Platforms come into being only through an already existing critical mass of users and data. This requires a complex set of sub-level networks that underpin the platform. And these networks, in turn, depend on an interoperable set of technical standards and protocols to already be in place. According to Langley and Leyshon, "platforms are particular comings together of code and commerce: when infrastructures of participation and connectivity are designed and data is realized and acted upon, this is the intermediation of digital economic circulation in action."[10] Nick Srnicek describes platforms as digital infrastructures that enable two or more groups to interact.[11]

Older technical definitions of platform may sound like this: "The entire hardware and software context in which a program runs. A program written in a platform-dependent language might break if you change any of the following: machine, operating system, libraries, compiler, or system configuration." In this description, originating from the Perl programming community, we see the meaning of interdependency at work: take out one element and the system no longer works. Unfortunately, such tech definitions lack the necessary neo-liberal rhetoric. They would not survive the economic onslaught of the managerial consultancy.

For business, the platform becomes a meta-product, a novel architecture for capitalization. US management gurus McAfee and Brynjolfsson, for example, sell platforms as a synthesis of social flows with a socio-economic intention. The duo defines platforms as "a digital environment with near-zero marginal cost of access, duplication, and distribution."[12] In reality, platforms are Hegelian economic war machines that control the user experience. They aim to subordinate users, firms, and indeed anyone involved in the making of products, to its economic logic. "A ride across town is a product, while Uber is the platform people use to access it." Get used to it. The sovereignty of production is about to be broken, turning all producers into suppliers (Amazon is the perfect example here). How are we going to make public the huge damage to society that the great cost reducers are causing?

In the subtitle of their book, McAfee and Brynjolfsson use the defensive military rhetoric of "harnessing our digital future" to describe the platform condition. Harness against what? The pre-internet "deep market"? Attacks from the networked multitudes or the Chinese and Russian secret services and their hacker armies? We're getting close… In fact it is openness itself they fear most, the open network that has led us to "malware, cybercrime and cyberwarfare, darknets for exchanging child pornography, identity theft and other developments that can make one despair for humanity."[13] Needless to say, platforms are not playing any role in these developments.

In terms of capitalization, it is the generated data from exchanges that really elevates platforms to a new level. Platform owners scrape off value—at least until an unforeseen disruption makes the ecology fall into pieces. The nature of this collapse is unknown. Backed by massive economic interests who will strive to keep these platforms afloat, we can only speculate that their demise will be different from the way platform pioneers such as Friendster, Hyves and Bebo turned into ghost towns overnight. What's certain is their vulnerability when it comes to information complexity. So far, we only know a fraction about how platforms come into being; soon we will find out more about how they decline.

BENJAMIN BRATTON'S THE STACK

The Stack by Benjamin Bratton appeared in 2016. This monumental cosmology, comparable to Peter Sloterdijk's three volumes on spheres, probes designs for a planetary megastructure in a truly cold, Hegelian fashion. Bratton's book should be read from an interdisciplinary perspective. When it came out, I read it as the successor of Lev Manovich's *The Language of New Media* from 2001.[14]

According to Bratton, networks, media and platforms no longer exist as separate entities. Earlier understandings that tried to read the world via the media angle have been rendered inadequate. Instead, "the stack" implies that everything is layered and connected, integrated into a larger digital infrastructure. While not explicitly discussed in the book, the stack has a technical origin and is still used by computer engineers when referring to the different layers of network architecture. One recent example is a layered post-Snowden visualization of "the internet is broken" that moves from the ground level hardware, operating systems and routers all the way up to user interfaces and apps, indicating that activists are working on all levels to fix it. For Bratton, the "planetary megastructure" of the stack has six different layers: earth, cloud, city, address, interface and user. As with any strong concept, once you start looking for stacks around you, they are everywhere. Whether television or healthcare, agriculture or logistics, it all uses the same cables, software and providers. This is the "infrastructural turn", a direction in which many designers, architects and scholars are working, from Keller

Easterling, Nicole Starosielski and Metahaven to Orit Halpern, John Durham Peters and Ned Rossiter.

The ecology of the geek mind is a self-contained entity that presents its imagined technological reality as a fact. In a similar fashion, Benjamin Bratton presupposes globalization of IT infrastructure as a given. "Planetary computation" as he calls it, is our a-priori, something given or assumed. This is the new normal. And yet, at least historically speaking, the cloud is a state of exception, a recent anomaly. But let's not create false opposites where there are none. Bratton thinks in pharmacological terms. The stack is cure and toxin, norm and exception, it is "powerful and dangerous, both remedy and poison, a utopian and dystopian machine at once."[15] Rather than either-or, the stack is both-and.

In the same way, the stack is neither a pseudo- nor a supra-state. Bratton rightly locates the essence of the matter in the contentious relationship between the two. In line with Carl Schmitt's tradition, we could call *The Stack* a political theology. The state is described "as a kind of machine, a vast apparatus for which the instrumental rationality of inputs and outputs should guarantee predetermined outcomes. Platforms, however, feed on the indeterminacy of outcomes." They are not machines. Platforms "have much more varied relationships to nonstate forms of authority and non-capitalist economies."

According to Bratton, the cloud has gained an independent status, becoming a fourth estate alongside land, air and sea. These classic three entities, once analyzed by Carl Schmitt in *The Nomos of the Earth* (1950) now have a new brother or sister named cyberspace. Debates have been raging for decades whether this "sphere", as Peter Sloterdijk termed it, functions as an additional layer or should be recognized as an autonomous world brain. Now that we've entered the infrastructure era, it is time to bring the four elements together into one overall analysis of "space force". What's at stake here is the question of who owns the internet. What interests lie behind this contemporary version of the Roman road system?

Defining the "nomos of the cloud" is the real contribution of the book. The stack is an (open system) architecture that Bratton has expanded to a geopolitical level. Although living in Southern California, Bratton is far removed from the surfer dude, aka New Age hippie. Nevertheless, there's an echo of Kevin Kelly's "inevitable" here. Regardless of anyone's opinion, the stack will happen. It is tempting to go along with the technological realism on display here. Bratton demands that the stack come into existence—and this is what all interesting theory should aim for.

However, knowing that regressive realpolitik might destroy the cyber episteme, Bratton presents his proposal in a Machiavellian fashion, concealing his intentions. We see Bratton acting as an advisor to power (even though he lacks the tactical playfulness à la Sun Tzu). This role is not unheard of in the world of architecture, but is relatively new in new media design. Bratton-the-designer constantly runs several scenarios simultaneously. The world may remain under US military hegemony, or it may not. Google may become a truly sovereign player, or it may just evolve into another brutalist-boring US corporation. The world may be divided, Cold War style, into clearly defined geopolitical territories through a delicate balance of multi-polar centers; or alternatively it may become really messy, relying on administration through Western globalist protocols.

My own reading of Bratton's bet is a technocratic future that will no longer be run by geopolitical entities such as China or the USA, but by a technical universal grid. "As computational edges and nodes claim some authority by their programmed automation, they also possess more authority as decision-making shifts from the designer to the designed." In line with Paul Virilio, Bratton warns: "The platform sovereignties that emerge generate their own unplanned productive accidents."[16]

Regardless of which future scenario occurs, each is only one of many possible worlds. And it is this ambiguity that is the weakest part of Bratton's grand design. If it does exist, it can be falsified, questioned, measured and given a legal status. If it doesn't exist, nothing is lost. Certainly science fiction and speculative theory, from McLuhan and VR to robot love, have played an important role in the creative-subversive process of designers, artists, architects and programmers. Yet Bratton seems to bet on both options, backing proposal and reality, and this opportunism weakens his integrity. The approach is reminiscent of Putin's advisor, Vladislav Surkov, who funded players throughout the political spectrum with the aim to establish a state of confusion. In a similar fashion *The Stack* contains all possible worlds, comparable to Ashby's definition of cybernetics as the domain of all possible machines. He describes the space of all possible relations. It is a thinking that does not permit an outside. The stack as a metaphysical meta concept may never exist and can therefore easily be dismissed as yet another Californian male dream of world creation, aka domination. All that said, platforms do exist, and this simple reason makes it less easy to put Bratton aside.

Platform is not one of the six layers of the stack, as one might expect. In principle, platforms should hover somewhere between the layers of the cloud and the city, but in Bratton's design of the stack they don't. Right at the beginning of the book, he decided to give platforms a special status, clarifying the relation between stack and platform. Bratton defines the stack as a combination of platforms. While stacks are platforms, most platforms are not stacks. Because of the facilitating nature of standards, platforms set the stage for the actions of others. What's important in this context is Bratton's description of platforms as a standardized diagram: no standards, no platform, no stack. In this way the platform can "tactically glue together lots of different things at different scales into more manageable and valuable forms."[17] According to Bratton, "platforms pull things together into temporary higher-order aggregations."[18] He defines them as hybrids, organizational forms that are highly technical and feed on the indeterminacy of outcomes. As organizations, they can take on a powerful institutional role. Platforms resemble both markets and states without conforming completely to either. The result is a new form of sovereignty, a third institutional form.

Whereas platforms operate on a horizontal level, stacks are defined by vertical integration. For Bratton the stack is a "vast software/hardware formation, a proto megastructure built on criss-cross oceans, layered concrete and fiber optics, urban metal and fleshy fingers, abstract identities and oversubscribed national sovereignty."[19] What limits *The Stack*'s usefulness in public debates is the inability of the book to read scenarios as ideologies and present alternative blueprints. Frank Pasquale, for instance, makes the obvious division between two narratives of platform capitalism: the conventional and the counter narrative. Whereas conventional discourse claims that "platforms promote fairer labour markets by enabling lower-cost entry into these markets by service providers", the counter narrative asserts that "platforms entrench existing inequalities and promote precarity by reducing the bargaining power of workers and the stability of employment." While such dichotomies are too simplistic, they at least provide these scenarios with stakes, asking us to construct alternate designs and imagine preferable futures.

FROM SOCIAL MEDIA ALTERNATIVES TO STACKTIVISM

In contrast to Bratton, French philosopher Bernard Stiegler explicitly calls for alternative designs. For Bratton, the technical world is what it is. The question of what is to be done never emerges. From the French-European perspective of Stiegler, however, the quest is to come up with meta-concepts that can one day be used to build secure, decentralized, federated infrastructure able to counter these "entropic tendencies". These alternatives are envisioned and prototyped by a diverse network of initiatives such as IRI (connected to Centre Pompidou in Paris), the Plaine Commune research project (out of St. Denis, together with Orange and the University of Compiegne), the Ars Industrialis network and the pharmakon.fr website, the summer schools in Epineuil, The Next Leap and Digital Studies proposals to the European Commission and international collaborations in Ecuador and China. (We at the Amsterdam-based Institute of Network Cultures are also part of this ecology.)

Both Stiegler and Bratton stress the importance of localization, but differ in their judgments as to whether locality should be a legal entity tied to *nomos*. What both share is the focus on planetary scale computing. In *The Neganthropocene,* Bernard Stiegler included his introductory remarks to his 2017 summer school entitled *Five Theses after Schmitt and Bratton,* in which he emphasized the importance of the "local slowing down of the increase of entropy."[20] Platforms speed up rather than slowing down, increasing entropy "through the network effect and its self-referential consequences." In the fight against computational reductionism and the decline in linguistic value, we need to preserve "incalculable fields" that are "irreducible to averages". Such singularities and exceptions work against the principle of leverage that dominates the world of algorithms, artificial stupidity and automation, the destructive logic "inciting an immense and dangerous *ressentiment.*" To counter this destruction, Stiegler proposes the concept of "local integrity" as a way to undermine the globally scaled technologies that short-circuit deliberative processes. In terms of software, this would mean regional crypto-networks that would protect both individuals and communities through decentralized data storage. "Only a reconceptualization of data architectures, and, more generally, of the architectonics that constitutes the computational episteme of capitalism, will open up a path that could lead us out of what has already been called the Trumpocene."

At the end of *Platform Capitalism,* Nick Srnicek makes a political call to collectivize the platform. "Rather than just regulating corporate platforms, efforts could be made to create public platforms—platforms owned and controlled by the people, independent of the state surveillance apparatus."[21] He demands that the state invest in such public utilities. This requires nothing less than a revolutionary change in policy, a radical shift away from neo-liberal privatization and austerity. "More radically," Srnicek continues, "we can push for post-capitalist platforms that make use of the data collected by these platforms in order to distribute resources." Despite these claims, the platform logic itself is never questioned. This is the accelerationist perspective, a position that sharply contrasts with the federated grass-roots approach that intends to disassemble power itself into smaller units that voluntary collaborate with other cooperatives.

In an attempt to politicize Bratton's fuzzy definitions, hacktivists have divided the totalizing concept of the stack into three distinct spheres: private, state and public.[22] The private stack, at least currently, is designed for consumers, makes use of closed tech and is run by market principles. The state stack, on the other hand, has subjects, uses closed technology and is administrated by the state. And the public stack, owned and operated by citizens, runs on open tech and is governed by commons principles. Whereas some aspects are debatable and in theory could be different, in practice this is often not the case. The private stack could be secure, free and open—but in the end isn't (think of Google's control of the so-called open source Android operating system for smart phones). The state stack could run on open tech—but often doesn't. If only the state was there for its citizens... And what about this "public" label? Is it just a hidden synonym for state-owned? What if we renamed the public stack into the common stack? (See Chapter 10 for more on this term.)

Let's move on and define stacktivism as infrastructural activism that is aware of multiple interconnected layers.[23] This is hacktivism with a holistic awareness of the multiple levels that exist above and below "code". Stacktivism thinks (and acts and hacks and intervenes) vertically. While work on social media is no doubt important, as in the Unlike Us network for example, these architectures have reached maturity, a stability that closes down possibilities. Alternatives should be more than do-good copies. The stack-in-the-making is an ongoing project, open to problematization and speculation. Each of the elements—both individually and together as a stack—have to be designed, politicized, federated and occupied. And each of these levels has its politics, players and cultures. Stacktivism is a form of direct hacktivism that does not wait for the state or international bodies to come up with platform regulation.

Stacktivism promotes an integral, total approach that goes beyond geopolitical realism: "We do not want a piece of the cake, we want the entire bloody bakery." In order to get there, a reassessment of the cloud will be necessary. Clouds are material symbols of a centralized platform society that needs to be dismantled, similar to the peace movement's demand during the outgoing Cold War. We cannot preach peer-to-peer solutions (including cryptocurrencies) while remaining silent about the scale issue of big data. To campaign for a renaissance of decentralization is one approach, but it also requires a critical "infra" part. Stacktivism helps us to think from the outside all the way to the inside. Indeed, rather than just having a nice time offline (which happens regardless), we need to delve further inside: tracing "dark patterns", unveiling the politics behind AI, opening the black boxes, and demanding open standards. And last but not least, let's make stacktivism post-colonial so that it can deconstruct the Anglo-Saxon/Western bias of so-called global knowledge databases such as Google Scholar (in comparison to the pirate logic of Library Genesis).

Before we drift off into the tech delusion of global control, it's worth remembering once more the radical modesty of Adorno's phrase, *das Ganze ist das Unwahre* (the whole is the untrue).[24] How can we develop forms of protocological politics that aim at the core, yet operate with strong roots in daily lives, with strong ties in the localities that matter to us? Another key question for the next couple of years will be how we should relate to the platform-as-form. Finally, rather than diversion and fragmentation, the platform offers the possibility of synthesizing social energies and placing history making back into our own hands. The lure of platform power is definitely there. However, very few of us will be able to run a platform that is both successful and sustainable. The wish to centralize and scale up comes with a price. Who will we trust with this extraordinary responsibility?

The platform desire will have to be counter-balanced with new forms of network cultures. There is no federation, decentralization or tech sovereignty without living networks. Can we reinvent the network-as-form without a trace of cyber-nostalgia? And if the network is too loose, what form will replace it? And what would it mean to give up the planetary protocol level and go multi-polar? Are we afraid of the "balkanization" of the Internet, which has been a reality anyway for decades with geo-blocking and national firewalls? Who's going to argue against platform regionalism? What happens when we choose to intensify social bonds in our vicinity? Are networks of fearless cities actually a weak proposition, a sign of a movement in retreat? Will networks be able to distinguish themselves sufficiently from nationalism, identity politics and other forms of provincialism? The horizons of the world are wide and call to us, even as we acknowledge the "lost spirit of capitalism".[25] We choose a level, pick our fight, and act where we're destined to act. Terminological awareness can help us to make connections between different struggles and establish a techno-culture of solidarity and respect that has yet to be invented.

6

From Registration to Extermination, On Technological Violence

"Q: Why do we act like machines? A: We do not." S. Alexander Reed—"Hacktivism is not always about breaking into a system, sometimes it's about breaking out of it." Anon—"The vile pogroms of 1940's were by-products of the industrial revolution. Today's pogroms are by-products of the digital revolution." Max Keiser—"It's time we start to have a life" autonomous slogan, early 1980s— "Calling them Deplorables is euphemizing them. Maybe better to euthanize?" (tweet).

Ever since the Second World War and the following Cold War, the (networked) computer has been anything but an innocent tool. We know from its history that the Internet has deep roots in the military-industrial complex. Mass violence and genocide are elements in the social media debate that so far have been ignored, if not actively repressed; digital technology as violence. Such a thought was taboo, depressing and discouraging the internet activists and geeks who are badly needed for digital interventions in this contested space. Who else is going to build the creative commons, the free and open source alternatives, and the public stack, if the master's tools are essentially evil and beyond repair?

Is it justified to speak of information technologies as a "social atomic bomb"? In 2018, evidence surfaced that the introduction of Facebook and the genocide of the Rohingya were intrinsically linked.[1] The step from the t-shirt slogan of "Stop Being Poor" to "Stop Being" is smaller than anyone of us would dare to think. Computers administrate everything, including the procedures of life and death. In these extraordinary times of the alt-right and right-wing populism, it is no longer business as usual. It may not be enough to call for a decolonization of information technology. We need to think the unthinkable—also in our strategic digital domain. In his 2018 *The Message is Murder*, Jonathan Beller is precisely doing this:

Today's codifications, abstractions and machines, far from being value-neutral, are rather racial formations, sex-gender formations, and national formations—in short, formations of violence. Digital culture is built on and out of material and epistemological forms of racial capitalism, colonialism, imperialism and permanent war. This violence is literally inscribed in machine architectures and bodies.[2]

In *Homo Deus*, Yuval Noah Harari writes about the concept of surplus population as "the useless class" and what to do with them. As the blurb on the back cover reads: "War is obsolete. You are more likely to commit suicide than be killed in conflict." According to Harari, "we are on the brink of a momentous revolution. Humans are in danger of losing their economic value because intelligence is decoupling them from consciousness."[3] His comparison between the twentieth-century logic of genocide and population control and the twenty-first century condition of the internalized subject is crucial here. "In the twenty-first century the individual is more likely to disintegrate gently from within than to be brutally crushed from without." The twentieth century was the age of the masses. Yet, as Harari states, "the age of the masses may be over, and with it the age of mass medicine."

Homo Deus states that, as human soldiers and workers give way to algorithms, some elites may conclude that there is "no point in providing health for masses of useless poor people, and it is far more sensible to focus on upgrading a handful of superhumans beyond the norm." In this Piketty age, income disparity is growing exponentially.

Therefore "in the future we may see real gaps in physical and cognitive abilities opening between an upgraded upper class and the rest of society." The State and the elite may lose interest in providing this increasingly marginalized poor with healthcare. Harari warns of the "creation of a new superhuman caste that will abandon its liberal roots and treat normal humans no better than nineteenth-century Europeans treated Africans." However, there will not be a simple return of the same. "Whereas Hitler and his ilk planned to create superhumans by means of selective breeding and ethnic cleansing, twenty-first century techno-humanism hopes to reach that goal far more peacefully, with the help of genetic engineering, nanotechnology and brain-computer interfaces."

FROM REGISTRATION TO EXTERMINATION

From early on, the computer has been associated with population control and genocide. For me, this story started when I was age ten. Rather than just through radio, newspapers or TV, I followed the Dutch protests against the 1970 census closely because my mother had joined the protest movement. She had been a courier during the Nazi occupation as a teenager, transporting clandestine newspapers and false food stamps on her bicycle in the southern town of Breda. My grandparents, for their part, hid two Jewish ladies in their chemist and postal agency. Each time I ventured down the narrow stairs of my grandparents' food cellar, I was reminded that this cramped, dark space used to be their hideout.

Around the same time, I watched the first television documentaries on World War II, the German occupation of the Netherlands and the Holocaust. Monuments and commemorations of resistance to these events defined my childhood. The Polish tank that liberated Breda in September 1944 stood proudly in the nearby park. It was during that time too, that I first heard of the Nazi sequence of registration, counting, selecting, razzia, transportation and ultimately extermination. The 1943 dynamite attack on the public records office in the Plantage Kerklaan in Amsterdam by the group centered around sculptor Gerrit van der Veen always spoke to my imagination.[4] The memorial on café De Plantage is still there. Van der Veen's group also produced some 80,000 false identity papers. Most members of the group were executed in the dunes outside of Haarlem.

The idea of the anti-census movement was simple but radical: the German slogan "*Wehrt den Anfängen*" (stop the beginnings) was applied to the bureaucratic policy of counting the population itself. Possible good intentions had to be ignored. The census had to be prevented, even if that meant civil disobedience. The fact that percentage-wise, most Jews were deported and killed in the "civilized" Netherlands was already widely known in the 1960s. The combination of precise census data, a well-organized administration and the willingness of police and other authorities to collaborate with the Nazi occupying forces, who executed their genocide in a strictly legal fashion, had turned post-occupation Netherlands into a deeply divided country, a struggle between those who collaborated with the Nazi regime, those who claimed to be in the resistance—and a large group of opportunistic bystanders.

In this age of large ID databases, requirements everywhere to login via Facebook and Google, not to mention fingerprinting and facial recognition, resisting registration appears impossible. There is no social media without profiles. Everyone is tracked. Where could we possibly start? There are those that fight legal battles like *Europe against Facebook*, some that tamper with their data entries in the hope of confusing the machine, and a few who pay in cash, refusing to own a credit or debit card. But here I want to introduce another approach that is usually dismissed as Luddite: what if the computer is not a neutral tool and becomes structurally violent and turns against us? Such a view runs directly counter to the paradigm of the PC as a liberation tool for individuals and communities that emerged in the late 1970s. Nevertheless, it is a view that was once widely spread, reaching its apex, ironically, in the year 1984.

LIFE IS SABOTAGE

To recover this older imaginary of computation, we return to another moment in my biography, the publishing of Detlef Hartmann's *The Alternative: Life as Sabotage, on the Crisis of Technological Violence* from 1981, a German autonomist classic. This post-utopian document of my generation, awkwardly interposed between hippies and yuppies, disco and punk, has never been translated into English. No one is using the phrase "technological violence", just as no one is still referring to West Germany. There is good reason for this. The book, filled with Marxist jargon, breathes the bitterness of its time. Think of the ugly brutalist high-rises, the *Raststätten* and *Autobahnen* that I frequented as a hitch-hiker, Kraftwerk and Fehlfarben, the Red Army Faction, and Peter-Paul Zahl, a period visualized in the TV detective series *Tatort*. In terms of theory, this is a world in which Adorno had been replaced by Michael Ende and Michel Foucault. In this grey post-war reality, Hartmann is just one of many harsh 1968 critics, a cadre of academics operating outside the academy. Similar heroes of mine at the time were Konkret and Edition Tiamat authors such as Wolfgang Pohrt, Eike Geisel and Hendrik Broder.

For the vitalist Marxist Hartmann, life is not sacrifice. Its meaning lies in the act of sabotage. What remains of our human qualities is the will to rebel. His central argument is that humans are not machines. This a-priori does not grow out of some innate superiority, or a nostalgic or sentimental humanism. Neither does it stem from the ideals of the select few, the superhumans that hover above our petty concerns. In this sense, it shares no similarities with leftist readings of Nietzsche. Instead, for Hartmann, life remains an unpredictable factor on the periphery, a set of forces that often disturbs processes and thus needs to be controlled and tamed, if not altogether erased. "Life has become sabotage, precisely because it is life."[5] Humans are defined by Hartmann and his German autonomist generation as a useless residue, a futile dream, a non-productive remainder that refuses to be utilized, quantified and optimized. This leftover core he calls "subjective strangeness": the non-value unable to be measured, incorporated and exploited. From a bureaucratic perspective, this worthless remainder is threatening for the entire system and cannot be ignored. It thus needs to be removed i.e. exterminated.

In line with late 1970s West-German brutalist reality, we're subjected to the violence of institutions, from shopping malls to schools, hospitals to jails (today we would add platforms to the list). These pedagogical institutions all follow the logic of the machine that intends to correct, educate and "better" humans through control systems. No matter how they are implemented, their architectures are identical, adhering to the primitive logic of the machine. For Hartmann, the richness of human language, games and feelings cannot be captured by the poverty of machines. Hartmann's vocabulary is clear yet bitter, working in a similar direction to the early Foucault, yet different from his later biopolitics term which, after Foucault's death in 1984, blurred into a general category that captured everything and nothing, effectively depoliticizing his work.

For 1970s German autonomists, politics was not about life as such. They're not existentialists. Life needs to be defended against a very specific type of population politics that attacks specific groups in society with a cold, efficient, bureaucratic rationality. This is a specific reading of German fascism. In their archival ventures, the Hamburg School (as I call them) bumped into historical evidence that had to be uncovered. Starting from the lessons provided by Hannah Arendt, they started to dig deeper into the anti-Semite eugenics of the Nazis, both inside the Reich and in occupied Eastern Europe and Russia, the proposed "breadbasket" for Germany that had to be depopulated first before it could be colonized and then fully inhabited by the Aryans, assisted by docile slaves (Slavic people). Those that did not fit genetically, had to be deported and exterminated.

Forty years after being published, Hartmann's "life as sabotage", much like Agamben's "bare life" notion, offers a different perspective on our contemporary condition. Yet our conditions are also distinct. The result is a kind of tension between conforming to a norm and the creativity and productivity promised by the outlier. These days, corporations embrace difference by reducing life to a set of identities. Life itself, with all its oddities and anomalies, has become a prime source of capitalist exploitation.

Data are extracted from tiny differences in taste, consumer behavior and opinions, then run through various computational procedures, visualized and sold to the highest bidder.

According to Hartmann, the machine is neither progress, nor a necessary evil, nor even some monstrosity born from the human mind. Instead, the machine is defined by its violence against life—and this is not some accidental side effect. Reading Hartmann, I interpret the machine as a vector, a vitalist force. As he puts it, the machine is a strategy of violence, destruction, power and expropriation. Such a view corresponds to the cybernetic logic that we deal with in the context of the internet. Liberated from its liberal political correctness, life is no longer defined by victimhood. Life equals revolt, it is an uprising, a demand for freedom, autonomy and the verdancy of subjectivity. To put it in 1970s terms: computers staple and punch you—but we fight back. The emptiness that institutions produce can be flipped into a revolutionary subjectivity. The result is a "rage against the machine", a defense of the human remainder that Hartmann calls the "technological class struggle". For Hartmann, such a technological struggle has always been at the core of the class struggle. In retrospect we can read this materialist analysis as an avant-garde statement. And yet it feels like we're still at the very beginning of this process.

Take the understanding of subculture, for instance, as an outcast element that capitalism can no longer absorb. For nearly half a century we have only seen subcultures from a productivist angle. In the highly cynical but dominant reading, every outcry, no matter how disturbing and unusual, can and will be sublimated and integrated into the capitalist machine. The notion of "the underground" is an impossibility; in capitalist realism, there is no outside. From the moment it is instigated, every subversive gesture has already been disarmed, every resistive act integrated into the machine. This stops us from acting altogether and makes us depressed. For Hartmann, on the other hand, these interventions are productive as long as they explicitly unravel power without compromise. Resistance is not futile, but fertile.

ITALIAN AND GERMAN VISIONS ON AUTONOMY

The German branch of autonomists stay close to the Italian school of Tronti, Bologna and Negri that reached its heights as a movement in 1977. Detlef Hartmann was part of the scene that published the radical theory zine *Autonomie*, with its subtitle "Materials against the Factory Society". Even though its discourse was highly abstract and sophisticated and included footnotes and references, the magazine did not adhere to the academic journal format, instead taking the form of the DIN A4 political brochure. Issue #1 reported on the 1979 Iranian revolution in support of the Shia opposition and Bani-Sadr, while #2 focused on new prisons and the trials against German armed struggle. The third issue from 1980 dealt with "The Second Destruction of Germany": the desolate concrete suburbs, the pedagogy of social housing attempting to tame the working class, and the strategies of squatter movements. In all these struggles, autonomy was a central motive, an aim to create independent lives that hold off and undermine the Machine (including the Party) and its imperative (always brimming with good intentions) of rationality.

Issue #13 from 1983, "Imperialism in the Metropoles: The Technological Attack", which unfortunately doesn't name its authors, is particularly relevant for our context. Its topic is the datafication of society. The aim of the editorial collective was to capture an overall picture, making sense of recent, seemingly disparate developments such as microelectronics, computerization, the introduction of robots to car manufacturing, data transport, and the rollout of cable television in major cities. Interestingly, the collective warns about overemphasizing the role of computers in surveillance and repression, instead stressing that computation sits within a broader regime of racist population policies and atomizing class struggles. The magazine starts off with a long historical overview that culminates in a new scientific phase of sabotage, which they describe as "computer guerrilla". In line with their Italian comrades, the collective displays a strong interest in the working conditions within post-Fordist auto manufacturing plants. Case studies on Volkswagen and Alfa Romeo factories demonstrate that the "operaist" dialectics between class struggle and automation are leveraged strategically by management in order to get rid of rebellious elements amongst the labor force.

Central to Issue #13 is a text titled "First Hypotheses on Information Technologies as a New Stage of Class Struggle." It describes capital's attempts to produce work without a working class. The point here is not that production entirely dismisses living labor, but rather that capital seeks a mode of "social-less" production that refuses any collective incorporation of the psyche. The data—not the workers—becomes social. This is reflected in the rise of post-Keynesian mass unemployment and precarious labor conditions. To sum up, what defines this historical process is the destruction of class-as-such. Within this development, the *Autonomie* collective has a special interest in the automation of social policy instruments that the late welfare state developed to control its (unemployed) work force, along with the computerized administration of the poor who are simultaneously atomized and centrally controlled.

IBM AND THE MERCILESS CAPTURE

The Hamburg doctor Karl-Heinz Roth was (and still is) a central figure in the German autonomist Left. In 1983–4, I was living in a West Berlin squat. Hans Sharoun's *Staatsbibliothek* (state library) had just been built right next to the Wall, on the ruins of the vast Potsdamer Square area. Seated amongst the brand-new stacks, it was here I eagerly read *Die restlose Erfassung* (The Merciless Capture), which Roth wrote together with the now famous historian Götz Aly. The book is a short yet significant historical study, detailing the census and the role of statistics during the Nazi period. It is here that I read for the first time about the widespread use of IBM's punch card technology by the Nazis in their 1933 census, its use within the military-industrial complex under Todt and Speer to coordinate forced labor, and its broader role within the Holocaust in terms of counting and selecting Jews.

Roth and Aly's book was written to support the census boycott movement, which had just celebrated a rare victory in a German court. Statistics were not just created to process large groups; Nazi tactics aimed to individualize, to isolate and extract single cases out of large databases. A blend of the scientific and rational, these methods sought to take out "subjective" elements of the social struggle.[6] But while Aly and Roth first sketch out the historical role of statistics during the Nazi period, they also showcase the individual careers of statistics experts that span well into the post-war period of West-Germany. Indeed, the authors point out that, while writing the book, the same punch cards were still in use.[7] *Die restlose Erfassung* thus contained a chilling but clear political message: collecting population statistics in order to single out social groups had a frightening historical continuity.

Although occasionally mentioned,[8] IBM's punch card technology and its role in the Holocaust is by no stretch a significant part of current Internet discourse. When Edwin Black's monumental study on the "strategic alliance between Nazi Germany and America's most powerful corporation" came out in 2001 (in the midst of the dotcom crash and 9/11), it received some airplay but couldn't be well positioned in the light of the internet revolution that unfolded at the time. The symbolic 50-year war commemorations were over and IBM seemed increasingly irrelevant, a relic of an old guard that had lost out against the newer baby-boom giants like Microsoft. In the popular imagination, the story of the punch card has also been sidelined for technical reasons. The punch cards were seen as a dead technology, relegated to the dusty historical archive of calculation devices. The computer arose in its place, a smart device that could do much more than merely calculate numbers.

From the very beginning, computers were presented as "thinking machines". It was Alan Turing's work in the 1930s that departed from the utilitarian adding of numbers to consider instead the concept of "general computation". The cracking of the Enigma code, in which Turing was involved, is widely known as the "crypto" World War II story. After this victory, the history of universal computation moves from John von Neumann through to the cybernetics of Norbert Wiener and Vannevar Bush's 1945 "As We May Think" essay (which marks the birth of the Internet). The computer as a calculation device recedes into the background, disappearing altogether with the rise of the PC in the 1980s. Computers could do more than put a man on the moon: they could answer all possible questions and process all types of information.

In his introduction, Black stresses that Hitler's "obsession with Jewish destruction was hardly original. But for the first time in history, an anti-Semite had automation on his side."[9] For Black, IBM's indifference to this racist remit was due to a blind belief in their own swirling universe of technological possibilities: "IBM was self-gripped by a special amoral corporate mantra: if it *can* be done, it *should* be done."[10] More than 2,000 card-sorting machines were dispatched throughout Germany, and thousands more throughout Europe. The so-called Dehomag Hollerith machines were deployed in numerous concentration camps. The "automation of human destruction" was conducted with the conscious knowledge of the IBM New York headquarters up until the last days of the Third Reich, with IBM subsidiaries custom designing the applications. A substantial amount of IBM's profit came from Hitler's Germany. Indeed, right up to the moment of America's entrance into the war in 1941, IBM's CEO Thomas Watson continued to proclaim his mantra of "World Peace through World Trade", circumventing trade boycotts through subsidiaries in third countries.

Retrospectively, IBM's direct involvement in the build-up to the German war machine between 1939 and 1940 presents a confronting piece of evidence. "IBM had almost single-handedly brought modern warfare into the information age. Through its persistent, aggressive, unfaltering efforts, IBM virtually put the 'blitz' into the krieg for Nazi Germany. Simply put, IBM organized the organizers of Hitler's war."[11] German orders not only included the production and maintenance of the machines, but also the printing of billions of electrically sensitive cards. As Black notes, far from being remote, generic deployments, each implementation of the punch-card technology meant extensive research and customization. "Each time IBM subsidiary field engineers had to undertake invasive studies of the subject being measured, often on-site. Was it people? Was it cattle? Was it airplane engines? Was it pension payments? Was it slave labor? Different data gathering and card layouts were required for each type of application."[12]

As a "solutions" company, IBM worked for both sides of the conflict. By 1943, two-thirds of IBM's U.S. factory capacity had shifted from tabulators to munitions. In the midst of this "merely technical" environment, one investigator working for the US Department of Justice, Harold J. Carter, did care. Between 1942 and 1943, he collected evidence against IBM in a document entitled "Control in Business Machines". While these investigations would ultimately prove fruitless, Carter learned that IBM was not just supplying the U.S. war economy and the army with machines, it was at the heart of nearly 100 military research projects, "including ballistics trajectory studies, aircraft design, automated inventory control systems, and an advanced wireless, electronic messaging unit called Radiotype."[13] IBM was also involved in some of the Allies' most top-secret operations, such as the Enigma code cracking at Bletchley Park in England, where they used Hollerith machines.

To demonstrate the difference that these machines made, Black compares two countries, Holland and France. Whereas the registration machine in the Netherlands worked flawlessly, in France "Nazi forces were compelled to continue their random and haphazard round-ups." In Black's words: "Holland had a well entranced Hollerith infrastructure. France's punch card infrastructure was in complete disarray." And his conclusion can be backed-up by numbers· "Of an estimated 140,000 Dutch Jews, more than 107,000 were deported, and of those 102,000 were murdered—a death ratio of approximately 73 percent. Of an estimated 300–350,000 Jews living in France, both zones, about 85,000 were deported—of these barely 3,000 survived, a death ratio of 25 percent."[14]

Black starts with the role of punch-card technology in the organization of census before moving on to the selection and registration of Jews and other Nazi enemies. Aided by new affordances of capturing, sorting and sifting, Black emphasizes a new informational imaginary: no one would escape. "This was something new for mankind. Never before had so many people been identified so precisely, so silently, so quickly, and with such far-reaching consequences." With this relentless hunting in mind, he makes the link to the omnipresence of computers 60 years later: "The dawn of the Information Age began at the sunset of human decency."[15] And yet a mystery still haunted the survivors. After having studied so many eyewitnesses and historical studies, "most of them would confess that they never really understood the Holocaust process. Why did it happen? How could it happen? How were they selected? How did the Nazis get the names? They always had the names."[16]

Toward the end of the war, the efforts of IBM and the Allies shifted toward tracing and saving these machines. Crucial economic data for the post-war period was saved in this way. Black's conclusion is bitter: "For the Allies, IBM assistance came at a crucial point. But for the Jews of Europe it came too late. Millions of Jews would suffer the consequences of being identified and processed by IBM technologies."[17]

So far, Black's monumental work, and even the role of IBM in general in computation, has been sidelined in the history of cybernetics. In Thomas Rid's *Rise of the Machines, the Lost History of Cybernetics*, for example, IBM is hardly featured, let alone its role in the Holocaust. Needless to say, it also failed to feature in the 1990s *Wired* vision that dismissed the mainframe computer as a symbol of the crumbling centralist age. But the overlooking of IBM's role seems like a side effect of historical amnesia and techno-positivity, rather than intentional conspiracy. In our understanding of early cybernetics, the focus was constantly on the individual, the user and the feedback between man and machine. Yes, control was core, aimed at halting the inevitable trend toward disorder and entropy. However control was always done in the name of a higher principle, the enhancement of human behavior.

For power analyses that question such a view of cybernetics, the emphasis instead should be placed on the automation of administration. Friedrich Kittler's insistence on the computer as a calculation device points in this direction. The all-too-human spell of the interface leads us away from the fundamental core of the machine. Unfortunately this is also the case of most of the cybernetics monographs that appeared over the past decade. Influenced by the writings of Norbert Wiener, Bertrand Russell once posed the question: "Are Human Beings Necessary?" This question directs our attention to (justified) concerns of human judgment and work being left out, foregrounding the real threat of selection and ultimately genocide.

It is this question that links the Reich's use of Hollerith technology with more recent events such as the 1970 resistance against the census in the Netherlands and the German protests in 1983. For these protestors, the big data collection of personal IDs, matched with identifiers such as religion, political beliefs and ethnic background was unacceptable. Black stresses, "with few exceptions, every Dutch Jewish family dutifully picked up the questionnaires, filled them out completely, and filed with the nearest registration office. The uncanny compliance was based on traditional Dutch respect for laws and regulations, as well as a penalty for not registering," adding that "Jews also had understood that resistance was futile because their names had already long been innocently registered as 'Jewish' in numerous statistical and registration offices throughout the Netherlands."[18] For Black, it was calculation rather than ornamentation that singled one out. "It was not the outward visage of six gold points worn on the chest for all to see on the street, it was the 80 columns punched and sorted in a Hollerith facility that marked the Jews of Holland for deportation to concentration camps." With this regime of organization and identification in place, one Nazi official could already report by October 1941 that: "all German Jews are now in the bag."[19]

A SOCIOLOGY OF LISTS

It was around 1984 that I discovered lists as a sociological category. The fact that lists do not merely exist, but were a distinctive concept, a mode of power along the lines of Michel Foucault, a specific way to organize subjects and matters, was a real insight for me at the time. This was during an era when the long lines of people waiting in the street for a bakery or office, had all but disappeared. Such queues were associated with dysfunctional real existing socialism and collapsing Third World economies elsewhere.

Once categories are in place, lists are being made. Lists empower. Lists repress. Lists order. So what could be better than publishing a comprehensive study on lists? When I grew up in the early 1970s, the list was the Radio Veronica Top 40, a folded sheet of paper I used to pick up in a record shop on Stadionplein for free. Later on, lists became a piece of software, a small, simple, yet highly powerful internet tool. In 1993 I got to know the electronic mailing list or listserv running on "majordomo", which defined lists of subscribers. Internet lists turned out to play an important part in my life, most of all for nettime, the mailing list I founded in 1995 together with Pit Schultz, and which still exists today. The practice of organizing networks for debate culminated in the "mailman" domain called listcultures.org that the Institute of Network Cultures has been running for the past decade. Communities that use listcultures.org include: VideoVortex, Unlike Us and MoneyLab.

Should we conclude that civil disobedience means sabotaging list making altogether? Lists are not innocent. This fight was not just about opinions, convictions, prejudices, and ideologies. Take the toys from the authorities. By posing the question of what lists are all about, we're entering dark territory. Lists are not by definition useful—there is more to this topic than the shopping list or the to-do list. In his 1960 magnum opus *Crowds and Power*, Elias Canetti describes the various cultural techniques that rulers have used over the centuries to prevent a crowd turning into a dangerous, unpredictable mass. People are separated and placed one after the other, in a line. Following Canetti, we could say that lists are abstract lines, cues that wait to be processed. In contrast to the open crowd that swells and then suddenly disintegrates, the list is stable and fixed. Surprisingly few items can (and will be) taken on or off the list. The list is a symbol of hierarchy, power and stability.

As a symbol for rational order, the list prevents the atomized subjects from unwanted articulations of collective energies. The communal becomes isolated and itemized; the crowd is made manageable. The chaos has been overcome; now we just have to wait and see how the number crunching is progressing. The institution will eventually deal with each and every single item. A list is not dead information. Rather than residue, it is a potent, dense form of rule that shows us the power of organization, and the organization of power. The list is living evidence, a reminder of the technological violence that inhabits our cybernetic machines.

Authorities require us to be on the list—and that we obey their rules. Once we're captured by the spatial order of the list, we cannot jump ahead in line or simply leave. This is by far the most dangerous aspect. Once we're on, how do we get off? From a database perspective, the list as an "organized collection of data" is an established concept.[20] For officials and managers, lists condense knowledge, grouping it into a specific order (often alphabetic or numeric). Abstracted and made machine-readable, itemized and organized data are ready to be processed. Once we've entered database management systems, the list as such disappears and is transformed into tabs, numbers, entries, forms, or simply "data" as it's called these days. It is only in the database that data become relational—as part of a list, data can be related to other data. But this is tedious labor—a task which has been taken over by the computer and earlier calculation machines that have become operational since the early 1900s. In 2017 our Institute of Network Cultures published Kenneth Werbin's study on lists.[21] This Canadian study not only makes a powerful, and potentially deadly form of power visible, it takes us deep inside the cybernetic logic itself in which the order of information becomes a prerequisite to virtually any move we take in this computerized, networked society.

DATA COMMUNISM AFTER DIGITIZATION

Sabotage, in Andrew Culp's terminology, falls under the category of conspiratorial communism. It's the practical translation of saying no to those who tell us to accept the world as it is. For Culp, it is crucial to break with the "legion of noncommittal commentators who preach the moderation of the middle."[22] Instead, his aim is to cultivate a hatred for this world by developing "contrary terms that diverge from the joyous task of creation."[23] The old German punk phrase "Destroy what destroys you" is presented as an effective cure against depressive disinterest. This entails activism and intervention, but such interruptions cannot be an end unto themselves. Culp recognizes that breakdown has become an integral part of capitalism, from the concept of creative destruction to the monetary collapses and ecological crises. The question is therefore:

what interruption is revolutionary? If there is anything that needs to be disrupted, it is the cyber myth itself.[24] How could the sabotage logic be applied to our data reality? Could big data be reengineered for the concerns of the common people, offering a solution to the current existential crisis of liberal-leftist progressives? In a contribution for *The Guardian*, Evgeny Morozov concluded that such data populism has

> a genuine advantage, but only if it understands that the traditional progressive agenda, like everything else these days, has been utterly disrupted by digital technology. Instead of denying it, progressive populists should use the data debate as an opportunity to re-establish their relevance to the crucial economic debates of today.

In short, data and databases are not evil and should be explicitly politicized. How can we create public ownership of our data? Can data, the "oil of the twenty-first century", provide economic growth, create new jobs, and even be the magic vehicle to redistribute wealth? Or is this all just a false utopia?

This was the question hovering over a gathering in November 2016 of the German radical-left coalition with the Hegelian name *Ums Ganze* (Claiming it All/Going for Totality) in an auditorium building at the University of Hamburg.[25] It was here I ran into Sandro Mezzadra from Bologna, whose work on migration and logistics I admire. He didn't know the organizers either. Email communication in advance had been scarce. Maybe they were too busy; maybe they were on a data exchange diet? The meeting didn't seem to be secret. Over the past decade, a neo-Communist movement had been on the rise in Germany, especially amongst a young, well-educated milieu. Thanks to the accelerationists, tech was now part of their agenda.

In an audience of 400 young people, Sandro and I were the only ones with laptops, and there were zero smartphones in sight, an anthropological anomaly these days. After a while, I did spot a few—but these were switched off and remained untouched. Much has been written about the discipline of the German working class and this scene was impressive evidence that these young comrades were determined to overcome capitalist temptations. Neither was the audience dominated by identitarian activists. For both Sandro and myself, the event was a time machine experience, transporting us back to a period over 30 years ago when we both lived in West Germany and admired the high-level political debates, the level of intellectual rigor, and the philosophical rhetoric of the strategic discussions inside the autonomous movement. It made us wonder: are these next generation communists Luddites, geeks, or both? Are we dealing here with a new division between the offline public life (in which we take notes on paper, as they all did), while covertly continuing to use digital tools in our private lives?

The meeting was part of a larger digital productive forces debate. Marxist insiders will be familiar with this key term, which describes why capitalism seems so vital and energetic. According to Marx, it is the productive forces that make capitalism so revolutionary. Marx's admiration for the productive forces is a problem for the young, turbulent, and romantic German mind that, by its very nature, is skeptical of the destructive and repressive violence of the machine. How do we work through these ambivalent feelings that are driven by objective contradictions? *Ums Ganze* explicitly asks why the centuries old debates about technologies and the left need to be repeated, over and over again.

Many parts of the autonomist agenda remain contradictory. The paradoxes and inconsistencies are apparent. What does it mean to be thrown from the accelerationist admiration of the digital productive forces to a radical Luddite condemnation of everything digital within one sentence? We're in Germany, after all, and these comrades can hardly be accused of being either profoundly confused or ruthlessly pragmatic. The two contemporary realities are deeply dialectic, so enjoy your synthesis.

How can we give such a schizophrenic relation to technology a productive dimension?

We might have to introduce a strong and appealing language that describes what's actually happening in the digital realm—one that overruns the typical managerial talk (agile, sustainable, disruptive innovation). Take the example of the computer file. In the dominant view, this is merely a mix of data and code. What would happen if we shift our perspective and see files as an accumulation of labor? How much human labor does your phone or laptop contain? We tend not to think that way because Silicon Valley has taught us to think along the lines of the economy of the free. But what would happen if computer science, internet criticism and IT journalism were populated by the grandchildren of Harry Braverman and Ernest Mandel? Or if social media analysts were Gramsci pupils? What if we revamped critical theory, removing references to nineteenth and twentieth century figures? Certainly we might reconstruct the "grand disconnect" that happened between historical Marxism and digital technologies—but that's only interesting for historians. More importantly, what would happen if we were to again take up these vanished threads, developing a digital Marxism free of historical references? Rather than limiting itself to the critique of working conditions, such a trajectory would engage with the digital on its own terms, developing a deep understanding of today's digitized production processes us such.

The Hamburg gathering seriously examined the two options: to reform digital network infrastructures or destroy them. Get your smartphone out and choose. Do you switch it off, or take out your hammer and smash it? Or do you take out your screwdriver and open its hardware, deconstructing its architectures and reprogramming its circuits and scripts? While the whole world outside was nervously clicking, trying to figure out the meaning of Trump, the German perspective inside the auditorium turned out to be both detached and refreshingly radical. Why bother with the latest tweets and items on your Facebook Newsfeed if you can think through its underlying structures? This is a dialectic jump, shifting the debate away from the personal computer as a universal calculator and the smart phone as a communication coordinator and instead thinking about the broader digital machinery that is used to carry out work. How are we going to sabotage data centers? Is it sufficient to "zero day" attack them, or do they also need to literally be burnt down and destroyed? How can we democratize sophisticated info warfare hacker tools that are now only in the hands of a few intelligence services and specialist cyber security firms?

At the event I was told about an English translation of a political pamphlet by one of the speakers, the Cologne Capulcu collective (of which Detlef Hartman seems to be a member). The pamphlet is called "Disconnect—Keep the Future Unwritten."[26] In it Capulcu calls for digital self-defense, a "refusal to take part in the permanent digital transmission." Instead, the collective urge a "counter attack on the praxis and ideology of total acquisition." The brochure argues against "convenience, comfort and velocity." In their historical overview, they distinguish between inventions that can be useful and innovations that are seen as the "onset of a big cycle of reorganization and a renewal of capitalist command." Over the past two centuries, machines have been introduced to undermine the craft of workers, lower wages and take people out of work. This attack is far from over and Capulcu points at the next wave of unemployment ushered in through the deployment of algorithms, the Internet of Things, the blockchain and robotics. Of course, these developments have a history; there is a long tradition of Luddite resistance strategies and counter-strategies from scientific management to suppress them. What has changed is the perspective of ordinary people "that perceive Big Brother no longer as a threat but a reliable friend" (with, for instance, advertisement seen as suggestions). Given these conditions, Capulcu stresses that time is our ally. "Slowness interrupts their streams of data." Stop giving feedback. The collective's motto for resistance against a self-regulated information society? "We haven't lost, we just haven't won yet."

The pamphlet zooms in on the healthcare side of the "technological attack" and the cynical politics of selection via "medical creditworthiness". This attack is carried out in a highly personal way through approaches like the quantified self, in which smart phone apps, bracelets and smart watches are anchored around a model of behavioral economics, one that anticipates a "genome revolution" in which we're asked to give away our DNA. Ultimately this type of surveillance secures submission. The digital economy, the collective concludes, is a "violent, patriarchal technology, realizing the principle of optimization and exclusion that destroys the social."

In late 2017, Capulcu published a book on this topic entitled *Disrupt!*, which aimed to come up with a practical technology critique.[27] The collective asks why we are so docile, so devoted to the technosolutionist logic of Silicon Valley that claims to "solve problems we did not have in the first place." Who needs self-driving cars? Not drivers, but the data hungry corporations that want to control our movements. Your AI assistant will only assist if you give data in exchange. Capulcu analyzes the "innovation offensive" as the motor driving the perpetual cycle of bubble > stagnation > populism > war. We need to break out of the future. This is freedom in chains. The collective calls for an attack on self-optimization apps and nudging logic, with its subtle behavioral changes, which presume limited free will and attempt to game our decision-making process. We don't want to be nudged!

The radical left gathered in Hamburg could hardly be accused of techno-fetishism. The dialectical struggle is asynchronous. On the positive side of things, the digital has unveiled its true self, shape-shifting almost overnight from starry-eyed hype to a cold, almost imperceptible element of the capitalist accumulation process. Digital technology is seen as complicit with the neo-liberal project that has created unprecedented income inequality and environmental damage. On the negative side of things, a doom scenario is emerging in which technology seems to be creating a vast army of "surplus population". In the German analysis, precarious living conditions under surveillance capitalism are the last stage of the marginalized classes before their inevitable extermination in war and genocide. Lumpen proletarianization is the general trend around the globe. These roaming subjects are no longer (potential) factory workers. For Marx and many who came after him, the industrial proletariat was the embodiment of a progressive promise of capitalist productive forces. But this is no longer the case. Today, the entire world has become a potential (digital) factory—and the inability to define where the borders of this global factory lie defines our present uncertainty. What if a small handful of factories in China or Africa can produce all the material goods the planet needs? And what happens if those last industrial centers get hit by automation?

What seems utterly absurd for outsiders is a thrilling exercise for all those aliens from outer space among us ready to debate the most unlikely of all scenarios: what are we going to do after a revolution in Germany? This was seriously discussed in Hamburg, without any irony. If we were to develop a 100-day program, what infrastructure should be prioritized for takeover, what can be used, and what should be switched off immediately? One can have a good laugh about such naive romanticism, but this is what happens when you reach the upper limits of Hegelian thought and allow yourself the luxury (or necessity) of moving up to the level of the "totality". Slavoj Žižek has always understood the value of seeing the forest (whereas many other theorists and academics, including many in cultural studies, lost themselves in the trees of trends). A similarly ambitious proposal was the *Bilderverbot*—the prohibition of images. Again, why not negate the entire world and overcome the addictive eye candy of Instagram, YouTube, television and film in one radical move? Why bother studying the complex workings of memes if we can overcome such regressive visual culture in one go?

This all sounds liberating—at least for a brief moment. Relax, don't do it, stop searching for the revolutionary subject. The avant-garde doesn't have to be reinvented and can be suspended. Let's fast-forward through history; it is happening already. This is the accelerationist premise. The multitude and precariat won't have a heavy responsibility as an emerging class. A commons-based, planned economy can begin tomorrow. Sadly, the computer came too late for the Soviet Union and East Germany. But now that everyone is equipped with unlimited computational power, if the collapse is imminent, why not prepare for the takeover in our own timeframe?

For a good 40 years, Antonio Negri has played a key role in the autonomy debates, both in Italy, Germany and elsewhere. Yet reading his recent work doesn't give us many clues for how to deal with the digital attack. In *Assembly*, Michael Hardt and Toni Negri self-confidently assure us that "ever since industrial civilization was born, workers have had a much more intimate and internal knowledge of machines and machine systems than the capitalists and their managers ever could."[28] For Hardt and Negri, rather than an antagonistic relationship, "humans and machines are part of a mutually constituted social reality."[29] We should "recognize the nature of the machinic subjectivities and machinic assemblages that are forming." The Italo-American duo admits that digital technology is a double-edged sword. Nevertheless they see young people resisting capital and governing the commons they produce. Which forms of resistance belong to this "machinic realism", if we may call it that? How can life and sabotage merge into one everyday experience? If we take the technological violence thesis seriously—and I think we should—we have an urgent need to construct a comprehensive ethics and a connected political strategy that is then widely shared. Many will mentally freeze (if not collapse) under the weight of the double-bind reality of the digital present. The cognitive dissonance between technological violence and intensely personal social media usage may ultimately collapse (or alternatively, fade away as more pressing issues take over). Seen from this perspective, is the activist demand for algorithmic sovereignty a false Hegelian synthesis or a clever way out? Isn't the proposed solution of individuals managing their own data and identity nothing but a structural impossibility and thus a reformist dead-end street? Our current inability to respond to platform capitalism and its ultimate form of technological violence is not a result of fierce debates and clashing positions. On the one hand, as Hardt & Negri assert, "Today we must immerse ourselves into the heart of technologies and attempt to make them our own against the forces of domination that deploy technologies against us."[30] On the other hand, we have Capulcu's call of resistance against the digital attack. Oppose these two views, and suddenly one gets a clear picture of how the radical left has moved itself into incomprehensible positions existing in parallel realities that very few know about. The first thing we need to do is organize the debate, inform ourselves about the multiple options of militant struggles against platform nihilism, and put all the options on the table.

7

Narcissus Confirmed: Technologies of the Minimal Selfie

"I cannot believe in a God who wants to be praised all the time."
Nietzsche—"Killing joy as a world making project." blog motto—"In the
present state of our social and economic accounting, I find it impossible
to say where necessary personalization ends and unnecessary personalization begins." David Riesman—"does this selfie make me look like i
didn't receive enough attention in my formative years" @stephsstone—
"Things fall apart. There's nothing you can do. Let a smile be your
umbrella." Jim Hougan—"We do not believe in the world empowering
women. We believe in women empowering the world!" Vilein—"It is
never my appearance that surprises me, but more so the fact that I show
up at all." Gabrielle Stein.

The selfie craze, starting with MySpace and taking off after 2010 with
the iPhone's first front-facing camera, should be read as a possible
survival strategy under harsh neo-liberal circumstances, one striving to
produce and maintain an identity.[1] "On retaining the self in a dehuman-
izing society" is the subtitle of Bruno Bettelheim's *The Informed Heart*.
Bettelheim discusses the survival strategies in Nazi concentration camps
in which being itself was faced with "man's destruction by his society."
Looking into the future beyond the mass grave, Bettelheim summed up
his analytical endeavor: "The success or failure of any mass society will
depend on whether or not man so reshapes his personality that he can
modify the society into one that is truly human; into one where we are
not coerced by technology, but bend it to our human needs."[2] Selfies are
contemporary evidence that, as Bettelheim's final chapter suggests, "men
are not ants."

Instead of focusing on "the extreme situation" as Bettelheim did, we
can ask ourselves what autonomy might possibly mean in the age of
neo-liberal hegemony. Today, as the neologism of "marxsism" (narcism
+ marxism) indicates, we no longer orient ourselves around solitary

evaluation, but around communal self-appraisal. Selfies aim to infuse design with dignity. Is the selfie the expression of the "informed heart" in the networked digital age? Can we bend technology in such a way that "self design" can liberate itself from corporate and societal constraints?

We're no longer obsessed with the hidden, contradictory nature of man that is supposed to be located behind the smooth images of the self. We do know that whatever the New Age used to call self and inner life was actually injected from its social environment (through lectures, books, movies) and that the real self is merely a recipient. That's the reason behind the demise of psychoanalysis as a cultural method. The problem with the polarized nature vs. nurture debate is that it is impossible to underestimate the role technology is playing. There is no self/ie outside the smart phone and social media and there is no self when disconnected. When it comes to camera angles, there is nothing left to be uncovered, or deconstructed: key features are highlighted, lightning and background checked, duckface is out, mirrors are your friend, blurry is fine, work your angles, a sultry expression is a bonus. The passion of the image is no longer a mystery. What remains hidden is the logistics of the image, from the Like economy to the political economy of the cloud, from the codex of one's camera to the filters and compressions of software.

But first, how do we deal with the selfie phenomena beyond forced participation or moral accusation, and develop ways of seeing that integrate machine readable interpretations, a real post-digital view?

Both art history and pop culture experts tend to agree that the self-portrait and the selfie speak about different subjects to different audiences. "The self-portrait and the selfie are two separate, though at times overlapping, efforts at establishing and embellishing a definition of one's self."[3] Those both for and against the selfie often see it as a defensive impulse to locate and protect an "authentic looking" subject through self-portrayal. Its makers, especially during the rise of the phenomenon, were often described as self-obsessed and compulsive. According to Alicia Eler "the selfie is a mirror, an illusion of a mirror, an egotistical moment wrapped in time, and an embarrassing moment post-shave."[4] An example of such organized narcissism would be the 2015 book of 352 Kim Kardashian selfies, published under the title *Selfish*. The complex differences between the portrait and the selfie nevertheless dovetail together perfectly when the selfie as "a mode of conversation, inherently contextual and often ephemeral" proves to be the perfect marketing vehicle for art museums, institutions caught up in the Like economy and highly dependent on visitor ratings and participation. Such contextual readings give little away of the energies of either practice.

For Western cultural critics, the selfie epitomizes neo-liberal self-promotion. There is a constant pressure to perform, to show off, to be present. The selfie embodies the desperate attempt by the failed individual to show that she (or he) is still in the rat race. There is a pleading quality that says: "I am alive, don't forget me, look at me and think of me, next time you can do me a favor." The advice we get everywhere is to never apologize for selfies. But who's afraid of being mistaken for Narcissus? Herbert Marcuse promoted the comeback of Narcissus, and was criticized for this. "If we believe what we're told by mass media, selfies are narcissistic, a product of a self-absorbed populous, vanity rituals of the me-me-me generation. Selfies are made by people, mainly girls, lacking in self-confidence, seeking constant validation from their peer group and beyond."[5]

In this forced intimacy, the viewer is between the camera and the subject, wrapped within the arm of the image taker. Are they narcissists or are they negotiators? They are the ones that hold the camera now, they pose to be looked at; they perform and experiment. Still, they display the disastrous reality of morbid selfies, snaps taken during funerals or in concentration camps. Such "reality" is far more real to the audience behind computers than to the selfie maker who observes them on a small mobile phone display.

How can we make a diagnosis that does not reduce users to addicts, an analysis that recognizes our deep pathology but refuses to simply declare everyone sick? In this technologized society it is becoming harder and harder to defend the right to freely theorize. Let's defend this intellectual space and surpass the old dichotomies of cultural studies vs. the Frankfurt School by overcoming political correctness on either side. The same can be said of the reading of selfies as the mere "portrayal of bourgeois self-understanding",[6] yet another genre in a bourgeoisie tool of self-representation of photography, now finally reaching a desired large audience. As networked images, underpinned by social and informational architectures, selfies are anything but autonomous.

Selfies privilege the present. Christopher Lasch's 1979 *The Culture of Narcissism, American Life in An Age of Diminishing Expectations*, marked out this problem early on: "Emotionally shallow, fearful of intimacy, hypochondriacal, primed with pseudo-self-insight, indulging in sexual promiscuity, dreading old age and death, the new narcissist has lost interest in the future." Lasch explains that this attitude comes out of a mood of pessimism, characteristic to the mid-late 1970s, reflecting a general crisis of western culture. According to him, "narcissism refers to a weak, ungrounded, defensive, insecure, manipulative self." There is no *intended* indifference here. There also seems to be no interest in the future, and equally little in the past. In line with the tumult of the '70s era, this collapse of chronology has created a vacuum that constantly needs to be filled with the evidence of presence, a consolation of losing the sense of historical continuity. In the jostle of the attention economy, the race to ensure this present personality is competitive at the level of visibility. However it avoids direct, head-to-head competitions through its game-like settings—if someone's noticed and Liked your form, you've already won. Time has collapsed into the current moment, the space between a person and his mobile has shrunk, and any individuality has been compressed into the same generic self-portrait. Selfies are a perfect loop in the transient moment of now.[7]

The "selfie" is defined as a photo in the selfie form. In other words, it is already a repetition that defers to other selfie. It is also a data trace that thrives on hashtags and categories, and is thus the opposite of the singular image that expresses authenticity. These imitations are habitual shortcuts, automated expressions, compressed gestures, in short: visual signs, utilized in order to escape artistic pretense. The deictic selfie gesture is the message. All it does is demonstrate presence, not a particular mood or feeling and it always points with its hand, paradoxically enough, to the device of the mobile phone. Selfies may express mass conformity: "I fit into this format". But they are also an expression of Lasch's "minimal self", a broken subjectivity that plays with irony. The ego is no longer considered a work of art; the act of merely upholding some kind of dignity amongst millions is a strenuous effort. What are the small differences allowed within these social pressures of individuated mass culture?

In step with Lasch, we need a theory of this minimal self(ie). Selfhood has become a luxury, out of place in an age of impending austerity. People have lost confidence in the future and begun preparing for the worst. The result is an "emotional retreat from the long-term commitments that presuppose a stable, secure and orderly world"[8] and a beleaguered form of selfhood. As Lasch repeatedly stresses, narcissism should not be confused with selfishness and egoism, because it relies on others to play the audience (as Echo did). Instead, narcissism is defined by the confusion between the self and the non-self. The play of desire for union with the world is hardly a symbol of cultural decadence and national failure. The problem here is neither selfishness nor self-absorption, but rather our lack of awareness about the status of digital portrayals in the age of face-recognition software.[9] Indeed, even over the last few years, their status has shifted; selfies are no longer the symbol of decline they were once imagined to be. At its analytical best, the fashion reveals social media's hidden obsession with ID registration, and the necessity to repeat one's visual presence, over and over again, in countless variations. Moreover, selfies represent the full paradox of privacy of our contemporary age, in which—contrary to surveillance studies and its obsession with faciality—other metadata, constantly collected and quantified, can speak more than any face.

A more materialist treatment can entertain the notion of these gestures as potentially subversive mass photo practices. The selfie is a prime example of individuation, the process described by Gilbert Simondon that Bernard Stiegler so often refers to.[10] As a product of an apparatus (as defined by Vilém Flusser), the selfie negotiates the conflict between the psyche and the collective in a technological document that is neither authentic nor industrial, but rather digital.[11] In many instances, the smart phone is anthropomorphized and used as a displaced object. What's important to investigate further is the link between individuation and identification—on a mass scale. Given these images are used as photographic evidence in the bureaucratic process of identification, we should not be surprised that the self can and will be used as currency. With Stiegler, we can say that there is a cognitive and affective proletarianization or deskilling happening, an anamnestic knowledge of images, a process where the externalization of memory becomes hyper-industrial. The selfie is an integral part of this process.

What kind of faith does the online subject place in this kind of presentation? We are not in the terrain of any truth therapy or self-examination of conscience. We're not seeking spiritual direction or alignment while updating our statuses. Social media are not "techniques oriented towards the discovery and the formulation of the truth concerning oneself."[12] They are not tools to know thyself, but instead to control one's self—for better and worse. At least this is the stated anxiety of the teenage heavy users that Sherry Turkle deals with in her 2015 book *Reclaiming Conversation*.[13] The purpose of social media is not the transformation of the individual. Selfies do not tell us what is hidden inside the self, but remain at the surface level, as the inter-face between humans. Our attempts to read introspection into a selfie bounce off the media surface. The object watches us; the selfie watches back.

American political scientist Jodi Dean disagrees with the douche moralists that "dismiss selfies as yet another indication of a pervasive culture of narcissism." The aim is to read amorally, and make a next step in the development of a technologically informed self-hermeneutics. The selfie, while being a kind of next-level portraiture, can be read as the end product of the democratization of media, ending the scarcity stage of image making, a symbol of our nihilist age of overproduction. What's central is the temporality of the selfie. "It's not meant as a commemoration. It doesn't memorialize what we've done. It's a quick registration of what we're doing. On Twitter, Instagram, Grindr, Facebook and Snapchat, selfies flow past, a kind of ongoing people's fabrication of the now." This shifts the discussion from the level of representation and its place in the archive to that of real-time culture. Selfies can be read as a necessary proof-of-presence, not as evidence of electronic solitude, let alone a symptom of a personality disorder. They do not exemplify who we are, but rather show that we exist at this very moment. Selfies are existential moments in a technological time, a "temporal hallucination" in Roland Barthes' words.

The self-taken picture is one that does away with any need for assistance from proximate others, seeking instead the responses of absent or desired others. Jodi Dean described selfies as images without viewers in this sense,[14] much like Adilkno's concept is of sovereign media: broadcasting to one's self.[15] Dean perceives selfies as a "communist form of expression", a proper "marxsism". Instead of praising or condemning our superficial egos, Dean stresses the social intellect of the image, or its circulation value, as she calls it. Dean does not see an existential digital monad that leads us further and further into the empty essence of the Western self, but instead positions the selfie inside the social network of relationships. For Dean, the provocative term, communism, is a reference to the common(s), not to some repressive avant-garde party that imposes its political and economic will on the people. In the selfie context, common means something less than full collectivity, namely, a minimal common, or what used to be called a mass or crowd. Dean:

Multiple images of the same form, the selfie form, stream across our screens, like the people we might pass walking along a sidewalk or in a mall. When we upload selfies, we are always vaguely aware that someone, when it is least opportune, may take an image out of its context and use it to our disadvantage. But we make them anyway as part of a larger social practice that says a selfie isn't really of me; it's not about me as the subject of a photograph. It's my imitation of others and our imitation of each other. To consider the selfie as a singular image removed from the larger practice of sharing selfies is like approaching a magazine through one word in one issue.

Milanese political economist Alex Foti has less crypto-optimism than this, linking the selfie to an increasingly precarious existence that constantly needs additional media validation to keep individualized anxiety at bay. In an email interview he wrote:

> It is a culture of naked, desperate self-promotion out there. My picture exists on social media and so do I. The earlier practice of Hollywood stars of taking snapshots of themselves (what we did in photo booths before the advent of the cell phone) has extended to the precariat at large: we're all aspiring starlets neurotically manicuring our own image for commercial appeal. Selfiemania bespeaks of existential uncertainty: Who am I? Am I really there? Am I what my own image projects?[16]

For Foti, the selfie is first and foremost a tool for self-promotion.

> Selfies are increasingly taken to make other people aware or envious of the trip to Europe or Antarctica and thus project a marketable self. A precarious person is constantly on sale in what are spot markets for temp labor. Job hiring is increasingly based on the first impression of the produced self-image, which has become a sort of avatar of our abstract capacity for symbolic labour in the affective economy of social media. It's all about exuding fake happiness and self-satisfaction. Nobody takes selfies of oneself in a blue or angry mood; Snapchat filters force you to be playful and funny.

For the activist in Foti, there is always hope that the camera eye will be turned to the turbulent world outside. "One of the surest signs that a rebellion of the precariat is under way, either in Paris or Hong Kong, is the fact that thousands of vertical screens are aimed at the spectacle of the multitude rather than at your precarious self."

A few years ago, Croatian independent media theorist Ana Peraica took over the photography shop of her deceased father in the Roman coastal town of Split. Surrounded by flocks of tourists on a daily basis, Peraica found herself confronted with the topic of the selfie and has just finished a study on it. I asked her why and how she was utilizing the term, narcissism, in such a free way.[17] Ana:

> What I found interesting about the myth of Narcissus was that it was miscomprehended in terms of visual culture, as Narcissus never made a self-portrait, he did not need a stored image but an active self-reflection in the water, in which previous images are completely irrelevant, as it is based on the process.

Peraica points out that narcissism has been excluded from the American edition of the Diagnostic and Statistical Manual of Mental Disorders, appearing only as a symptom of psychopathic behavior. "Around the same time, the fake diagnosis of 'selfitis' (the 'disorder' of being addicted to making selfies) was introduced into the media sphere and was claimed to be recognized by American Psychiatrist Association (APA), which is, of course, not true."

Instead of claiming that people who take selfies either need to go and see a doctor, or prove to us how their liberated, balanced life style results from such "tools of self-examination", how might we frame these images otherwise? Ana: "I am of the opinion we have entered the third cultural phase of Narcissism, the one more closer to the original myth of Narcissus, speaking of a full self-abandonment, that was already visible in media art pieces, which were producing self-portraits on a participatory level. They showed that there was no fixed self, but selves that are exchanged. Boundaries and integrity of selves do not exist anymore—not because of selfies but because of a slow media-based deconstruction of personal needs and scenarios."

Everyone is only too aware that we are taking selfies in the post-Snowden era. Ana:

> We are surrounded by a strong surveillance system that is now running on voluntary self-exposure by which a Narcissus objectifies

itself to be reminded what is the subject again, a surveillance system formerly known as Echo. And in that process there are plenty of calls for help, fears exposed, showing again how photography and imaging are becoming very important for a culture unable to self-formulate differently, i.e. by speech or writing.

Ana concludes that selfies are not bad in themselves, but a consequence of many bad influences: the loss of communication, the decay in education, and the disappearance of text as we once knew it. We are becoming empathic machines, mechanically reacting to whatever happens with a smile and a LOL.

How do we go about critical selfie research? For the time being, the selfie has not gone out of fashion. Teenagers are not yet disgusted by the mechanical gestures of the selfie. In 2017 Peraica brought together her work in a book, published by our Institute of Network Cultures, entitled *Culture of the Selfie, Self-Representation in Contemporary Visual Culture,* in which she examined the role self-portraits play in art history and developed her own mirror theory used when taking selfies. The book was launched at the *Fear and Loathing of the Online Self* conference in Rome, organized by John Cabot University, Roma Tre and the Institute of Network Cultures,[18] with speakers such as Jodi Dean, Wendy Chun, and Gabriella Coleman, and on the European side, Marco Deseriis, Olga Goriunova, Ana Peraica and Franco Berardi. How do we get beyond the predictable split between the politically correct claim of empowerment (of young girls) and the nihilist critique of self-promotion and despair? Is the selfie the visual evidence of a cataclysm of the self, which, in our post-digital condition, filled with boredom and disgust yet incapable of revolt, turns against itself? Reading selfies as folk-art is not really a way out as it does not neutralize a growing moralism, one constantly offering up medical prescriptions and recipes for wellbeing.

Can we still talk about the narcissistic personality in the selfie context, as if it were an individualistic trade? Not really. The selfie is first and foremost a technological gesture, produced by a specific hardware condition (a smartphone with a built-in camera, always-on cell reception, selfie sticks), threading through photography software and dispositions of being. Part of the documenting of "the now" is the minimal lag between photo taken and photo circulating on platforms such as Snapchat, Twitter and Instagram. We cannot talk about the selfie, therefore, and remain silent about the invisible Like-economy behind it all, the billion dollar advertising industry, the ubiquity of facial recognition software and the burgeoning surveillance market of people's private data. We've learned to accept that the selfie is quantified, the smile is commodified, and the resulting value is traded behind the user's back.

8
Mask Design: Aesthetics of the Faceless

"There is more truth in the mask we wear, in the game we play, the fiction we obey and follow, than in what is concealed beneath the mask." Slavoj Žižek—"My secrets won't make you happier" Amalia Ulman—"Stop treating the internet like it's a different thing and start focusing on what you actually want your society to look like. We have to fix society, before we can fix the internet." Peter Sunde—"We may be decentralized and disagree on a lot of topics amongst ourselves, but operations are always carefully coordinated." Anonymous—"Insults from complete strangers. This is the true promise of social media." Neil—"How valuable is reputation if any idiot off the street can rate me?" #peeple—Social media or "how to turn our thoughts violently towards the present as it is" (Stuart Hall)—"Man is the master of contradictions." Thomas Mann—#Apply: The same boiling water that softens potatoes, hardens eggs —"We are unknown to ourselves—and with good reason." Friedrich Nietzsche.

Indiscriminate scraping and analysis of personal data by governments and corporations such as Google and Facebook have virtually eradicated the conditions for what was once a core value of internet culture: anonymity. As a more playful, innocent phase of cyberculture before the medium became mainstream, early net culture offered a range of possibilities along the anonymity spectrum, from pseudonyms and multiple identities in role model games to anonymous remailers. Needless to say there was—and is—no such thing as absolute anonymity, now or then: in the last instance, everyone is traceable. Anonymity isn't a purely technical issue, but a social contract signed with the sys admins and their contractors, and built on the techno-libertarian consensus that data will not be passed on to commercial or governmental parties. Ever since the Snowden revelations in 2013, this relationship has been shattered. From that moment onwards, as Michael Seeman observes, *Kontrollverlust*[1] became all too obvious—we have lost control and the social contract has been broken. We cannot possibly trust smartphone manufacturers, ISPs, platform operators or even crypto-software, let alone our anonymous brothers and sisters.

Instead of individual and defensive—protection against the intrusive policies of governments, corporations and fellow citizens—let's emphasize anonymity as an offensive gesture of collective performativity.[2] Rather than a mere technical means to save what is left of our privacy and personal autonomy, anonymity here refers to digital forms of impersonal heteronomy that go beyond the fear of being touched in order to embrace "the joy and thrill of engaging in ephemeral encounters with unknown others, of momentarily suspending one's 'real' life in dissimulative role-playing, and of losing oneself in a proliferation of digital masks."[3]

Anonymity works best when it is understood as an act of playful masking. It's all about the process of transformation and becoming. We may read the longing for anonymity as a European romantic gesture,[4] but anthropology provides us with different readings. Take Roger Caillois' 1935 essay "Mimicry and Legendary Psychasthenia"[5] or Claude Lévi-Strauss, who both urge us to look at what a mask transforms and excludes, not merely what it represents.[6] In *Crowd and Power*, Elias Canetti uses the example of archaic figures to expand on this idea of metamorphosis as a kind of oscillation between the human and the animal. "It has often been pointed out," he writes, "how fluid the world was then. Not only could a man transform himself into anything, but he also had the power to transform others."[7] Yet if the mask is a means of metamorphosis, its powers are specific and singular.

It can be distinguished from other forms of transformation by its rigidity. Unlike the many changes a face can make, the mask is fixed and fossilized. The mask separates, and is supposed to reveal nothing of what is behind it. However, the actor can always wear a second mask underneath the first one.

The power of unmasking is part of this story. This is where paranoia comes into play. For the paranoid, Canetti tells us, "the wealth of appearances comes to mean nothing; all variety is suspect." The paranoid "has the gift of seeing through appearances and knows exactly what is behind them. He tears the mask from every face and what he then finds is always essentially the same enemy."[8] The winning strategy here is counterintuitive: even if there's ample necessity for a wide range of people to take privacy measures, anonymity as a culture really only thrives when individuals temporarily come together, merge into a swarm and go public.

Paradoxically then, anonymity works best if it's part of an offensive strategy, practiced by many, out in the open. As soon as it retreats back into the legal realm, becoming a right or an obligation, it loses its magical, transformative energy.

Gabriella Coleman has written the epic history of the Anonymous movement. In it she explains the background of their signature icon, the white plastic mask with black stripes. She reminds us that

the Guy Fawkes mask was a pop cultural icon thanks to the Hollywood blockbuster *V for Vendetta*. The movie portrays a lone anarchist's fight against a dystopian, Orwellian state. The mask had also appeared previously on 4chan worn by a beloved meme character with a penchant for failure—Epic Fail Guy.[9]

The specific history of the 2008–12 Anonymous movement, eloquently described by Coleman and captured in the documentary *We Are Legion*,[10] is a tragic case of the slogan "United We Stand, Divided We Fall". Anonymous brought thousands to the street with their slogan "We Are the Internet", sided with Wikileaks, broke up with Julian Assange (like most of his supporters), broke down after betrayals and repression, and then dispersed, only reappearing occasionally. Anonymity is a serendipitous state of becoming—like most political events these days, we never know if and when it will manifest itself. Alongside anonymity are interventions such as the Distributed Denial-of-Service (DDoS) attack or the company hack. Rather than articulating dissent within traditional institutions or striving to effect societal change, these new forms of digital disobedience aim to be "a direct intervention in the networks of control and economic circulation that govern the present system."[11] All we can do, then, is to be prepared when it is time to swarm, ready to use all the means available while remaining aware that there is no such thing as absolute security, impenetrable privacy or the perfect mask.

In her 2015 epilogue for the second edition, Coleman summarized the lessons learnt for social movements. "Anonymous has clearly enabled a new political position, one where *actions* matter, and actions can be evaluated, but the identities behind them—even when they are identifiable and subject to prison—are acknowledged by all involved to be less important than the actions they perform." And yet as Coleman stresses, the coordination and planning of this "anonymous" action is not carried out in secrecy. "These activists organize on public chat channels, issue press releases, and announce their causes and offer reasoning in dramatic videos. They are typically in direct contact with local non-Anonymous activists and journalists."[12] The entire episode of the movement can thus be summarized in this one sentence: "Belief in the idea of Anonymous is enough to motivate action, even if full anonymity is not the goal or is unachievable."[13]

Collective anonymity can be developed and articulated in many ways. Take the imaginary theorist and aphorism expert Johan or Johanna Sjerpstra, presumably a Dutch sociologist, whose name is used in various countries and circumstances whenever a quote requires an author. The history of collective anonymity is a rich, ever expanding story. The yippies, mail artists and situationists, Black Mask, Against the Wall Motherfucker and other Neoists all used similar tactics. We might also remember Coleman Healy, Karen Eliot and Monty Cantsin (an "open pop star" invented by Al Ackerman in 1978). Other examples of anonymous collectives are 0100101110101101.ORG, Adilkno (Bilwet), Sonja Bruenzel, Tiqqun and The Invisible Committee. These experiments range from collective identities whose names are not necessarily secret to individual pseudonyms such as Hakim Bey or Tinkebell.

One tactic is to use an (imaginary) person's name for a movement of many unknown actors, an approach found on the fringes of conceptual and performance art from the previous decades. The most well known of these is probably the Luther Blissett Project, the Italian collective behind "Q". Members of this group had already been heavily involved in the *autonomia* movement in the seventies and eighties as well as operating the Mao-Dadaist radio station Radio Alice. As Luther Blissett, the collective pushed this idea further, coining the term *con-dividuum*:

> It is necessary to get rid of the concept of In-dividuum, once and for all. That concept is deeply reactionary, anthropocentric and forever associated with such concepts as originality and copyright. Instead, we ought to embrace the idea of a Con-dividuum, i.e., a multiple singularity whose unfolding entails new definitions of "responsibility" and "will," and is no good for lawyers and judges.[14,15]

In 2000, a group from Luther Blissett started the writers collective Wu Ming. The Chinese term can mean either anonymous, unknown or five people, and also refers to the third sentence from the Tao Te Ching: "Unnamed is the origin of Heaven and Earth." Working together, Wu

Ming has now written several novels such as *54* (2002), *Manituana* (2007) and *Altai* (2009).

Luther Blissett's strategy differs from that of The Invisible Committee—while the former is an imaginary author, the latter is a collective that speaks with one voice. In 2007 the anonymous French collective released a pamphlet called *The Coming Insurrection*, encouraging readers to "flee visibility" and to "turn anonymity into an offensive position." What do such movements teach us about visualization and personification? Luther Blissett is a mask, but in movements like Anonymous the mask itself is the mask. And what does all this say about the relationship between the individual and the collective? Are we seeing the creation of an escape route for the codified subject, a search for larger aggregates? Such overarching identities seem to be seductive, inviting us to be assimilated and thereby transformed. But is it better to design imaginary entities yourself or to connect to a larger collective?

In the midst of all these questions, it's important to remember that anonymity revolves around a game of hidden identity at a specific moment and in a specific context. Anonymity is above all a temporary experience, a ramshackle structure that works as long as it works and may disappear as soon as actors feel they're no longer receiving a return on their investment. Dogmatically sticking to Temporary Common Denominators is stupid—it's better to pass them on. That's what the internet taught us.

The *Faceless* exhibition that premiered at the Amsterdam art space Mediamatic surveyed the role anonymity plays in contemporary design and media art.[16] Artist Bogomir Doringer, who curated both the online and offline incarnations of the show, related his interest in anonymity to the emergence of facelessness. During his study at the Rietveld Academy, he noticed artists and fashion designers engaging with anonymity as an experimental aesthetic and political form: using masks and other cloaking devices to mutate the natural shape of the human face, to morph between official and perceived identities, and to more broadly explore new surveillance technologies and their paranoiac forms of perception. As Doringer observes: "The unstable identity of the present begs for the return of power of the mask from ancient times, when it was used as a form of protection, disguise, performance, or just plain entertainment."[17] The return of the mask was exemplified by the work of Carmen Schabracq, in which woolen masks and sculptures foregrounded the intimate association between anonymity, the body, and animality so well-established in the pop culture of the preceding decades.[18]

Two related responses to anonymity occur in the work of photographers Frank Schallmaier and Hester Scheurwater, who both take as their research object the online self-presentations of gay men and women. Schallmaier collects selfies and penis-comparisons on gay dating sites, presenting them in a thematically organized manner.[19] Scheurwater, for her part, is inspired by the anonymizing (in)formalities of selfies, the way in which imagery and the complicit observer work to render the body transparent. Here "identity" is not constructed through a face-name pairing, but through the materiality of the universal body. Centered around the recording mobile device, this atypical alignment of legs and arms without a face recalls the alien life forms described in sci-fi novels. First published on the social media platforms they drew from, Scheurwater has since organized these images into a booklet called "Shooting Back".[20]

Also on view at *Faceless* was "Islamic Carding" by Iranian artist Shahram Entkhabi, a single image that juxtaposed two seemingly irreconcilable aesthetic and ethical regimes: Western culture and Islamic culture. For Entkhabi, the former is afraid of anything that refuses to show itself; the latter of what might be revealed when it happens. The critique of the burqa is typically articulated in terms of women's rights, violated by a backward religious doctrine. Yet—without wanting to replace this explanation—an alternative hypothesis for explaining our unease with the burqa-phenomenon is possible. The burqa frustrates the Western imperative to render objects transparent, to make them structurally available for incessant circulation and subsequent consumption. Ironically, this economic transformation "gives them away" in a similar manner to the giving away of women in so-called backward cultures.[21]

An aesthetic approach to anonymity as a form of (dis)identity politics is the polar opposite of a legalistic defense of privacy. Artworks engage with the politics of surveillance and the politics of (post) identity. As part of a movement-in-the-making, they broaden our imagination to include new forms of collective life based on sociality, forms that embrace exposure but nevertheless defy the super-panoptical machinery. Premodern connotations and functions of the artist resurface. This is the revenge of the jester or trickster, where deceitful appearances invoke a temporary retreat from historical seriousness and enter the playful zone of subversion.

The sad part of online culture is the presumed singularity of the self, an exhausting play with minor differences in the name of self-promotion. This is a game that many hate, yet few know how to escape. While the liberal focus on the individual's right to privacy may prop up our feelings of dignity, it simultaneously narrows down our options. Exhibit A: Facebook's war on multiple identities. As Patricia de Vries writes, instead of "fortifying a stronghold for identity and self", we could instead open a "wider playing field that welcomes relational multiplicities."[22] Understanding the self as interdependent and relational helps us to multiply the exits out of our data-obsessed prisons. We are, by default, embedded, comingled, hybridized and stained. Each of us is the sum of volatile and contingent relations and mediations, of countless alignments between disparate actors, settings, and things. Take a so-called individual and pull any loose thread; she unravels into a web of myriad interconnections and affective relations, a tangled mesh winding from the micro to the macro level. When you trace these connections, an infinite map of relations unfolds, some ambiguous, others harmonious, some strong, others weak, some contradictory and others competitive.[23]

In her essay "Dazzles, Decoys and Deities", Patricia de Vries suggests,

> we are in need of movements that resist the temptations of a binary universe in favor of emancipating, productive, affective and relational forms of critique. We need critics that resist branches of neo-positivism, aestheticism, and individualism. Over and against a binary thinking of neat demarcations and isolated domains, thinking in terms of relations opens pathways to intersectional forms of critique.

This raises the question of how to give form to a self as a sum of ever-changing relations. She writes: "This is a strategy of affirmative presence, as a process of imagining otherwise, of making worlds, constantly opening the doors to unknown futures outside of algorithmic tracking wars and calculated settings."[24] Reframed as something larger than the singular self, identity here might prove less amenable to individual capture, commodification and control.

It is not hard to observe that privacy is a bourgeois commodity one can purchase in exclusive places, accessible to the affluent classes. The tactical withdrawal of the rich into temporary white spots—a vacation on Mars—is only one of many responses. The 1% has a whole array of both human and non-human services at its disposal, allowing them to quickly move through public space without being noticed. The options for privacy online are far more limited. All you can do is install crypto or malware protection software, add browser plug-ins, or use VPNs and secure email providers. Hacking or being hacked is today's equivalent of eat or being eaten. The global poor cannot afford privacy and pays the price, providing free immaterial labor to the data giants. Needless to say, surveillance capitalism reproduces poverty and marginalization. The latest paradigm is "offline as the new luxury". The happy few can delegate their communication-and-control to their personal assistants. Meanwhile, the cognitive precariat is encouraged to be online 24/7, compelled to text and navigate their tribe-like social life through their smartphones, waving their devices up in the air to find a signal, struggling to coordinate their lives with productivity-enhancing apps, suffocating in traffic jams while commuting to and from their temp workplaces. Welcome to the digital divide 2.0.

We play with identities until it's time to tear off the mask. We scare, surprise and seduce, but eventually the veil has to be uncovered. The current preferred way out is offline, into the reality parks and tech-free zones without a signal. We dream of going incognito on real life streets, moving through the zones of invisible non-identity temporarily provided by facial concealment strategies. The mask project promises to remove the rails of our online playgrounds, to free us from virtual enclosures such as *Second Life* and *Call of Duty*. However the unplugging is not done à la *The Matrix*: the idea is not to return to a reality. To put it in vulgar Marxist terms, going offline for the weekend is the twenty-first-century version of the reproduction of labor power together with a festival visit, a yoga session and a good joint.

After Snowden, the question is no longer how to uncover the panoptic regime. The prison has moved its walls inside the self. That's the move from Foucault's centralized and institutionalized disciplinary society to Deleuze's societies of control. The Deleuzian call for "new weapons" to wield against the societies of control has been a never-ending source of inspiration for artists, designers and activists alike. But what comes after internalized control? Clueless about the next steps, we put out feelers, investigating in all sorts of possible directions. We're aware of an inner takeover, but what can we actually do about it?

The comfort of the smooth interface lulls us to sleep. Speculative aesthetic art practices have problematized the loss of privacy from different perspectives. Some mock the culture of paranoia, others soak it up. Some seek to undermine monitoring technologies by way of camouflage and invisibility. Others push against the ostensible immateriality of surveillance technologies, calling for hyper-transparency and making visible the invisible. And some even try to disrupt the calculative mechanisms of specific software, or warn us of an encroaching bleak future if things do not radically change.

Why was the internet so appealing? For some, it provided an ability to speak and act without others knowing who you were or what you represented. For others, it was a chance to build an entirely new personality for yourself or your group. Yet today, the commercialization and militarization of the web has led to the concatenation of online and offline identities, a synced, singular identity supported by the hegemony of client-server architecture. This development is reinforced by the type of bourgeois sensibility that shrinks back in horror at the messiness of the social. Anonymous communications are intrinsically risky. Better to retreat behind the safe walls of the Facebook community and cultivate one's "true self" within a select circle of family and friends. Trapped in this glass house, the citizen-user finds herself immobilized, falling back into the depression that has now become the natural psychic state of today's office temp.[25]

Snowden's revelations about data surveillance have further disenchanted those who continue to value privacy and anonymity, whether online or off. In today's socio-economic climate, unconnectedness only serves to render us obsolescent and opacity merely seems to incriminate us ("to attract secret services, please encrypt"). Such dead-end strategies drive old-time defenders of the socialist appropriation of mass media, resulting in technophobic grumblings reminiscent of Thoreau or the Unabomber.[26,27] But when it comes to imagining alternatives to an ever more omniscient network society, recourse is often taken to legalistic or ethical measures. We present unlimited technological possibilities with one hand or advocate a user license agreement and ethical code of conduct with the other. Where's the third option?

So the need for artistic and activist experiments with collective forms of anonymity remains. Contemporary society is camera-saturated. Yet it also seems bent on restricting aesthetic response, on containing any flourishing of the same singularities that it has helped unleashed by monitoring, commodifying and personalizing experience. In the face of these safeguards, anonymity unlocks a range of dangerous alternatives, releasing pure new forms of energy: "The question is how to re-imagine anonymity not as an attainable categorical state, but as a way to recoup an energy of metamorphosis, the desire to become someone else."[28] The artistic subversion of the projects discussed so far lies not in their use (or abuse) of internet applications, but in the conditions they create for open characters, in the broadest sense of that term. Consider the automated creation of user profiles; the invention of fictional whistleblowers; search engines where queries are lost rather than found; conspiring digital body parts inspired by Chatroulette; the automatic creation of bulk Facebook profiles by bots, the control of group behavior and the emergence of collective identities within MMOs. After casting off the dire self-management regimes of Facebook, there is a whole world to discover. These projects by artists and activists remind us that, in fact, we have to take away even more from the NSA revelations, to truly come to terms with this moment. Despite the legitimacy of a certain weariness with postmodern solutions to fundamental political and socio-economic problems, these types of identity games can short circuit an immersive post-Snowden defeatism.

The Guy Fawkes mask is by no means the only possible character that can be adopted. Global popular culture is a rich resource for what we could call "common denominator design". A particular mask design can start off as a meme, go viral across the globe within weeks and become a recognizable brand—a potent political signature understood and employed by billions. To better understand these dynamics, we need to distinguish them from the generally ephemeral meme, from the speed at which memes come and go. The question becomes one of culture. How might we assemble a lab where such experiments can be created and then tested out in a small community? One of the outstanding issues would be the (re)introduction of anonymity in popular social media platforms such as Whisper, an anonymous secrets sharing app (with issues) that launched in 2012.[29]

Michael Dieter once told me that there's something to say for a society where anonymity doesn't require wearing bizarre masks in public. The urge to don a mask should indeed be seen as a sign of crisis. The compulsion to hide is a moment that marks a society in transition. In today's society, the existing institutional forms of power and representation such as the political party, the trade union, the church or the tribe have all but disappeared. It is in these dire times, marked by permanent depression, that Anonymous has emerged.

And yet Anonymous was never merely an escape route for the disenchanted, but a call to team up and act together in public. Far from scaring us, the heterogeneity of its strategies should instead urge us to further theorize these new formations of the social. Anonymous is a collective artwork, a performance of the social potential of the precariat as a networked class. In overcoming the antinomies of bourgeois thought, it escapes, just for now, the unbearable truth of identity. As the "social movement" concept meanders into the dead-end of the NGO and the current network discourse is too weak to replace it, all we can do is extend feelers into the darkness, creating a never-ending stream of masks as collective identities.

The artistic projects discussed here broaden and reinvigorate our imagination, even if they remain ineffective at disaggregating the forces of capture and control. These artists realize that the paranoia endemic to mass-surveillance has now become a weakness. Protecting the individual-that-has-nothing-to-hide while targeting the terrorist is relentless, exhausting and ultimately impossible. Forever haunted by the existence of the unknown, these regimes will eventually collapse under the weight of their own energy-guzzling data parks. To amplify this paranoia toward its eventual breaking point, we dive into anonymity. This is the disappearing act, the leap into the unknown. The preferred tactic to accelerate this process is not to expose the status quo—the favored path of critical thinking—but to mask ourselves.

9

Memes as Strategy:
European Origins and Debates

"Those who laugh have not yet heard the bad news." Bertolt Brecht—
"Oh, plenty of hope, an infinite amount of hope, but not for us." Franz
Kafka—"Is his lack of fame justified?" Johan Sjerpstra—The nihilism of
data without carrier—"No Socialism for You" (meme)—Cybernetics of
the Deed.

CYBERNETICS OF THE DEED

If the mask is an activist strategy, so is the meme. But if the mask is
primarily concerned with protection, memes are all about production.
Memes symbolize the historical development of ever-tighter human-
machine bonds. Memes form a bridge between the mental and the medial.
In *Rise of the Machines*, Thomas Rid describes "the team" as one of the
core concepts of cybernetics: creating a single entity between human
and machine that behaves like a "servo-mechanism".[1] Memes are an ideal
object for understanding this paradigm. They are neither abstract, pure
ideas, nor a biological neural substance, but something in between; they
are PNGs and JPGs, but also addictive and viral. A meme can never exist
in and of itself; it needs to appeal to users, to be shared by them, and
to adapt itself for optimal spread through endless replication. Through
this circuit (or feedback loop, or recursion, all have computational
undertones) of human-machine interaction, a piece of visual content on
the internet becomes a cultural artifact—and a powerful one at that.

This cybernetic metaphor surrounding memes can be traced back to
their roots. British biologist Richard Dawkins coined the term, meme, in
his 1976 book *The Selfish Gene*. In it he describes tiny elements of culture
that travel from person to person through copying or imitation, adopting
various elements as they circulate. "Human brains," he claimed, "are the
computers through which memes operate." While the meme concept
was there, most of Dawkins' text is so outdated as to nullify much of

its etymological purpose. After all, Dawkins was writing pre-everything: pre-internet, pre-social media, pre-Pepe. For instance, he used memes to describe long-term changes of culture. But today's internet memes on boards and social media are transient and highly ephemeral, barely lasting a few hours. The content and context of a specific meme is topical and rapidly evaporates. What counts is replication, alteration, satire, and, above all: an ability to know your speed. In 2013 Dawkins made a decisive clarification, characterizing an internet meme as one deliberately altered by human creativity and thereby distinguishing it from his original idea involving mutation by random change and a form of Darwinian selection.[2]

To know your meme means to be a pop culture insider and share the cultural matrix in which the memes circulate. Today, the need to understand memes has become urgent, and no story better illustrates this than that of Morris Kolman, a former unpaid intern who worked under staffers as part of the 2016 Hillary Clinton presidential campaign. In his early twenties, he approached me with a long email exposé on the "political potentials and pitfalls of memes."[3] Seeing "a great kind of community building potential in memes," as part of Clinton's digital audience development team, he sought to integrate memes into the campaign's social media strategy, following the success of others like Bernie Sanders' "Dank Memes Stash". But in response he was simply told: "We don't do memes. The Internet doesn't like us." Whilst we all heard the conventional explanations for this—that Clinton's persona was, for example, too rehearsed—we might also understand their response as what has been referred to as "cruel optimism", a denial of what this young man calls "a suffering that is being lived through" that he sees as "a foundational aspect of meme culture."

In his undergraduate thesis on the topic, Kolman teases out the deep relationship between cybernetics and biology present in meme culture. Memes, he argues, are not only products of an increased connection between human and machine; they also display the cultural anxieties of that shift. Drawing from the work of WJT Mitchell, he figures memes as paradigmatic of Mitchell's "age of bio-cybernetic reproduction."[4] As technology has permeated through our lives further than in Benjamin's time, the images created by these technologies have taken on aspects of this new mode of existence. Memes spread, evolve, and replicate; the actor responsible for a meme's explosion can often be narrowed down no further than the ambiguous social-technological construction of the

Internet. Indeed, there are even memes about how much people communicate via memes.

For Kolman, the burgeoning popularity of memes over the past decade can be heavily attributed to the rise of "networked neo-liberalism", the integration of traditional neo-liberal subjects into an exponentially more productive and draining online milieu, where all production can be optimized and all identities can be visualized. One needs only to look at their Facebook advertising preferences to see themselves as a pile of fragmented character traits, a piecemeal personality. The conditions that this creates—a constant need for connection, a dependence on social affirmation, and a general malaise as more aspects of life and identity are outsourced to technological means, spiraling out of our control—all play themselves out in the egomania, hedonism, depression, and nihilism rampant in millennial meme culture.[5] Ironically, it is precisely this set of conditions that make memes such a good container for these affects. Memes proliferate (and find hospitable ground in social media) through their ability to create feelings of relatability and connection between users on the network. The survival of the memetically fittest breeds something highly efficient, easily digestible, and continuously perfected for socialization. Examined in this light, it should come as no surprise that memes have become a hotbed of online political culture.

But what political culture are memes a hotbed for? In the aftermath of the rise of alt-right, the meme debate has inevitably been narrowed down to a specific US-centric trajectory in internet culture, from 4Chan, Reddit and Know Your Meme to Milo, Bannon, Breitbart and Trump. This is well documented in Angela Nagle's *Kill All Normies*, in which she describes the trajectory from Obama's Hope sign to the racism debate surrounding the Harambe meme. Nagle concludes that "the online culture wars of recent years have become ugly beyond anything we could have possibly imagined." How did it happen that we come to associate memes exclusively with the digital counter-revolution? According to Nagle, the dominant memes are not so much evidence of a return of conservatism, but instead of the "hegemony of the culture of non-conformism, self-expression, transgression and irreverence for its own sake—an aesthetic that suits those who believe in nothing but the liberation of the individual and the id, whether they're on the left or right. The principle-free idea of counterculture did not go away; it has just become the style of the new right."[6] Memes as such do not come into being ideologically aligned one way or another. Instead, what we see here is that there are attributes of memes that the right has seized on and made great use of. However this shift is just the latest iteration in a longer history that, as we'll see, stretches back decades. By uncovering some European origins rather than the predominant American memetic narrative, this history contributes some fresh ideas on how to recuperate the meme.

THE 1996 MEMESIS DEBATE

Thomas Rid once remarked that Norbert Wiener, the founder of cybernetics, tended to anthropomorphize machines: "switches corresponded to synapses, wires to nerves, networks to nervous systems, sensors to eyes and ears, actuators to muscles. Conversely, he also mechanized man by using machine comparisons to understand human physiology."[7] For Rid, whereas cybernetic concepts like control and feedback were hard to grasp, the merger of human and machine easily stirred the imagination. The crux of this fascination turned on a deterministic promise: a sufficient understanding of the human as machine could lead to the operation of a human at the ease of a machine. Memes, these new cybernetic images, are no different.

Like many of my generation, I came across memes in 1996 when I was asked to moderate the online debate of the Austrian electronic arts festival Ars Electronica, curated by the newly appointed director Gerfried Stocker. He chose "the Future of Evolution" as the overarching theme, aiming to "identify palpable-memetic-conditions under which the cultural development is proceeding."[8] Twenty years after the invention of the term, memetic science was being popularized, brought back into the popular imagination by the rapid speed with which the digital revolution was recalibrating the world.

And yet lingering in the background of this theme was the continental European tendency to short-circuit debates about culture with those on biology and "human nature", embodied in phrases such as "cultural evolution". Implicit in such an idea was that history is not man-made, but rather contains an internal logic that remains unseen and unknown by us, the ignorant citizens. The artist-engineers that discover these laws at an early stage will become the new rulers, whereas the rest will have to obey. We cannot revolt against nature. Once our cultural techniques have become second nature, the subjects will have to wait patiently for evolution to evolve to the next stage. The irony—especially among this crowd—was palpable. In the words of a *Wired* piece by James Gardner during the same year, "I'm not initially attracted by the idea of my brain as a sort of dung heap in which the larvae of other people's ideas renew themselves, before sending out copies of themselves in an informational Diaspora. It does seem to rob my mind of its importance as both author and critic."[9]

Memes entered into the discourse with an unquestioned power, a perfect meme would be almost teleologically bound for success, given that its power would be able to penetrate and alter even the most stubborn minds. That these cybernetic units quickly took up the language of the virus is only appropriate; viruses, too, are neither alive nor dead, neither technological nor biological. In the 1996 catalogue, Sadie Plant characterized this viral power. It is not openly visible and confrontational, rather, it scales up and overwhelms its host—any attempt at a cure coming far too late. Viral power is not constructive, especially not coming out of the 1980s, when the computer virus, the retrovirus HIV, and the "cultural immunodeficiency virus" entered the stage. It took a while for the communication nets to emerge, but once this was done, in the mid-1990s, the natural ecology for the cultural virus was ready for seamless replication, at the speed of light.

Mark Dery, having recently published *Escape Velocity*, continued this thread of criticism in his contribution to the conference in Linz. Dery addressed Kevin Kelly's 1993 cyberculture classic, *Out of Control*. Packed with biological metaphors and technofuturism, Dery contrasted Kelly's "rise of the neo-biological civilization" with the anti-computer writings of the Unabomber Ted Kaczynski, who, curiously, had been embraced by many of the early internet entrepreneurs. Dery warned of a cyber-capitalist revolution masterminded by a technocratic elite. This cast doubt on the potential for leftist artist-engineer radicalism through memes, by noting whom they shared an object of interest with. "The digirati lend their radical libertarian economics the force of natural law by couching it in language of chaos theory and artificial life."[10] Memes were more than just an apolitical phenomenon up for grabs. Their strength lay precisely in their effortless, organic power. They justified the ends to which they worked for no other reason than they were destined to work in the first place. Putting this back into the lips of the technocratic elite, Dery pointed to Michael Rothschild's *Bionomics: The Inevitability of Capitalism*, which states that capitalism "is not an ism at all but a naturally occurring phenomenon." Strip memetic logic bare and you are left with the classic totalitarian credo: might makes right.

This debate was endemic to the conference. The official position was made clear in a statement by the director, Stocker: "Human evolution, characterized by our ability to process information, is fundamentally entwined with technological development. Complex tools and technologies are an integral part of our evolutionary 'fitness'. Genes that are not able to cope with this reality will not survive the next millennium." In another contribution, Tom Sherman downplayed the importance of the motif of evolution while stressing the desperation that really drove the economic imperative to connect.

> You don't hear a lot of evolutionary analogies in factory lunch rooms or college coffee houses these days. The talk is about survival and how tough a place the world has become. People are forming relationships with machines, not necessarily because they're attracted to machines, but because they are desperately trying to get connected with other people, particularly with those who can help them survive.

With connection becoming a necessity for survival, internet memes not only mimic this biological instinct, but—echoing the previous argument about their relatability—actively reify that necessity of technological connection.

Richard Barbrook voiced his critique of these "dodgy bio metaphors" such as the survival of the fittest memes and warned of an implicit return of social Darwinism.[11] In his contribution to the email list debate, Barbrook demonstrated how easy it was to debunk the positivist mysticism of memes. "Hypermedia" should help to actualize the earlier promises and exercise media freedom. He called memetics dud philosophy, bad science and reactionary politics. "The refusal to be duped by false promises of the memetic nirvana is an important step towards ensuring that hypermedia is used to improve the daily lives of everyone."[12] That is, the conception of memes set forth in the conference ought not be adopted. A fundamental reframing would be necessary to recuperate the problematic dynamics fueling their efficacy.

Stepping back, *New York Times* critic Douglas Rushkoff attempted to diagnose the debate. Having just published his first book entitled *Media Virus*, he noticed that the meme debate "quickly degenerated into a fearful premonition of neo-eugenic, civilization-wide fascism."

The future meme world painted by critics was dark and Hobbesian. "We seem to fear that, left to our own devices, we will rape and pillage one another." Back then, Rushkoff saw cyberspace as "a free marketplace where the best ideas, or most useful ones, tend to survive." According to Rushkoff, "social theorists are victims of a campaign of social pessimism that advocate vigilance above all else." In short, fingers could be pointed at both sides. The masses must be led by a benevolent elite and carefully monitored and analyzed through polls and other testing. For Rushkoff, the meme critics that praise top-down mind control essentially feared progress. "That's why so many well-spoken social theorists hate us pro-internet, Californian-style utopians." Without an alternative to the straightforward cybernetic paradigm, the memetic debate was destined to run in circles.

Looking back twenty years later, Rushkoff wrote to me:

When I decided to use the viral metaphor for the transmission of ideas, I wasn't even familiar with Dawkins. But then when I read him, it seemed to me that there were a few things missing. Memes are a great corollary to genes, for sure. But just as the genetic scientists of the period underestimated the importance of gene expression, Dawkins' model was underestimating the importance of meme expression. under what conditions do certain memes flourish and others don't?... Scientists that focus on memes underestimate the importance of the culture in which those memes are attempting to replicate. They see the figure—the meme—but not the ground, the culture.

Rushkoff remembers that in the early days memes primarily circulated in advertising agencies:

It meant all they had to do was work on crafting the best meme for it to go "viral". But that's not what I meant. I proposed that repressed cultural agendas will find expression through viruses. The potential has to be there, already. Trump, the meme, replicates—at least in part—because there was already a widespread white nationalist rage in America. That's what I meant by "cultural immune response"— which is the real operational factor in any viral spread.

The cultural soup of memes returns us once more to Kolman. Drawing upon Vilém Flusser's work, Kolman explores the dynamics of how memes interact with their cultural conditions to trigger the massive memetic movements we see happening. As content is the main way we project identity and interact with others online, content consumption becomes a critical point of social unification. "People no longer group themselves according to problems," Flusser states, "but rather according to technical images."[13] Kolman extends this point.

If people are grouped by content consumption instead of traditional interpersonal identity groups, then images change from objects of symbolic meaning to the prerequisite for meaning creation in the first place... the more technical images take on the role of social connector, the more they reinforce the importance of their distribution mechanisms to socialization.

Thus, when Rushkoff points to the strength alt-right memes drew from white nationalist sentiments in America, we can go back to the original work of Flusser and see why this might be the case: "Media form bundles that radiate from the centers, the senders. Bundles in Latin is fasces. The structure of a society governed by technical images is therefore fascist, not for any ideological reason but for technical reasons." So it is not just the agency of memes at work here, but rather, as Rushkoff says, the cultural soup in which they arise. And what kind of culture is it? An image-centric one perfectly suited to supporting the kinds of connections that memes excel at initiating.

The end of our first example, however, ought not end on a note of nihilism. Kolman makes the point that

given the inherent circuitry of any technical image, the content of them is always something of an illusion. "They are like the proverbial onion: layer after layer comes away, but when everything has been understood, explained, there's nothing left." The popularity of KnowYourMeme. com is a good case for this. The question one always approaches the site with is "what's the joke here and why is it spreading?" After that answer, there is little left to think about.

This explains why memes have historically worked for both sides of politics, and at their core are not necessarily fascistic. Depending on the culture they arise in, they may support any particular project.

The question then becomes what can be done to shift that culture away from the current pervasive fascism seen on the likes of 4chan. Even this, however, risks falling into the fantasies of control that have been playing out for decades. If there is a lesson to be gained it might be found by comparing the memetic fascination with a similar lauding of evolutionary might. In Ridley Scott's *Alien*, the remaining crew of Nostromos is interrogating Ash—a cyborg they dismembered after finding out he was instructed to let them die—as to how they may kill the deadly alien aboard their ship. "You can't," Ash retorts, "you still don't understand what you're dealing with, do you? The perfect organism. Its structural perfection is matched only by its hostility." "You admire it," responds one crewmember. "I admire its purity. A survivor... unclouded by conscience, remorse, or delusions of morality."

Sadie Plant, reflecting on her viral power essay 20 years later, raises a worry analogous to our anxiety over the Alien's biological perfection.

To still be [thinking about the possibilities offered by memes] now—maybe things should have moved on? And perhaps this means that the discourse was itself limited, a dead end. I say this mainly because I don't know what to think about memes today, the depressing Pepe and his generators. The 1990s work seems to me rather like a piece of conceptual art: it's great while it's happening, but where does it take us, what does it makes possible?

The American collective Critical Art Ensemble (CAE) also wrote an essay about memetics for the online debate in which they took apart the "nature as ideology" idea. Following Roland Barthes, they claim that "under authoritarian rule, the social realm is divided into the natural and unnatural (the perverse)." At first sight, nature looks moral and pure. When the rules change "the dark code of nature (survival of the fittest) is efficiently deployed and genocidal nihilism becomes an acceptable course of social action."

CAE asked: "Why do we want to open this Pandora's box yet again?" Twenty years later, the question is on the table again. What do we gain by saying that a message has gone viral? How many of the kids would even notice the biological metaphor here—and what disastrous consequences this once had, not so long ago.

DANCE TO THE TECHNO VIKING

The second case is the Techno Viking meme based on a four-minute video shot by artist Matthias Fritsch during the 2000 Fuckparade in

Berlin.[14] Here is an example of an authentic meme, and proof of our argument that memes do not have to be associated with the right. Wikipedia summarizes the video like this:

The camera is on a group of dancing people with a blue haired woman in front. A man stumbles into the scene grabbing the woman. A bare-chested man (known colloquially as the Techno Viking) enters the scene while turning to that man. He grabs him by the arms and the camera follows, showing the confrontation. The bare-chested man pushes the guy back in the direction he came. He looks at him sternly and then points his finger at him, ensuring he behaves. Then the camera follows the bare-chested man as the techno parade continues. Another observer comes from the back of the scene offering an inverted bottle of water to him. As the situation calms down, the bare-chested man starts to dance down Rosenthaler Straße to techno music.

Fifteen year later Matthias Fritsch produced a "crowdfunded mon-umentary" that perfectly preserves the innocent pre alt-right meme culture of the Web 2.0 era before the brutal years of monopoly platform capitalism.[15] What's striking is the original fascination of those inter-viewed (including the maker) with the viral potential of social media in its early days, an element that was later taken for granted: "Copy, transfer, combine. Dominant is what's spread the most." What was intriguing about Techno Viking, and turned it into a cult, was the uncertainty whether the street scene is real or staged. The timing was so perfect that it lead to the common phrase: "The Techno Viking doesn't dance to the music, the music dances to the Techno Viking."

The video really took off only in 2007 when YouTube and 4chan started to spread the video and renamed the video from the original Kneecam no.1 to the name of the protagonist. In this golden age of user generated content, people started to remix the video, reenacting chains of gestures that became iconic such as the inverted bottle and the leaflet shredding. The remixers copy the "pathos formula", as cultural theorist Marc Ries described in the film. There are also Techno Viking performances inside Grand Theft Auto and Minecraft. At some point Fritch started to collect all the related materials and turned it into an archive, now based at the Karlsruhe Design Academy, arguably turning Techno Viking into one of the best-documented memes.

From the history of the Techno Viking, we may take a key point of solace—memes resist efforts at top-down engineering. The quest for a meme science is still a dream. As one theorist asserted: "If I knew what it was that causes content to go viral, I would not tell you, and I would be a billionaire by now. If we knew what does and doesn't go viral, we could design viral content and the advertisement industry would shower you with money." The chaotic and democratic nature of the formation of internet memes and their spread means that their growth will always be somewhat horizontal. The fact that good memes spread far also means that good memes have had a history of vetting. For those who may wish to engineer authenticity, such vetting is inevitably encountered and has to be consistently passed. We are not fighting an uneven battle.

THEY SAY WE CAN'T MEME

We are not fighting an uneven battle, but we are losing. As soon as we understand resistance as organized interference, we can start doing counter-mapping, monitoring the silence, and bringing to light the hysterical realism that has been hidden for so long. As we can learn from Silicon Valley business gurus, disruption is enough to bring down vast systems because of their meaningless routines. We claim to blast lasting holes in the self-evident infrastructure of the everyday. This also brings the possibility of a revolution closer—an event that even the most dogmatic critics of the neo-liberal regime ruled out ages ago. However, as George Monbiot insists, this is based on our ability to tell stories.

Developed and tested over countless iterations, these narratives can then be condensed into memes. Memes, we've established, are empty shells waiting for content. The problem is that even when memes are not right leaning politically in their narrative, the story they spread is of the inevitability of capitalist realism. We need a new narrative to fill these memetic receptacles. The overall narrative will have to be robust, while remaining agile. The core message has to stay the same, no matter how much a meme is altered. For this occasion I emailed a few authors, asking them what they thought of the meme question.

I first asked my fellow Amsterdam-based media theorist Marc Tuters how we could start to acknowledge the attractive side of memes.

As much as the whole meme war thing seems the ultimate end point of Rancière's claim that "to identify politics with the exercise of, and struggle to possess, power is to do away with politics", we should also acknowledge that, from the perspective of (leftish *Guardian*-type) millennials, political memes actually seem to make politics fun for once. Because of Trump we're all focusing on the dank memes disrupting politics for the worse, but after Obama won it was a different story, so we shouldn't just forget that. Before the alt-right found their so-called meme-magic, the left had its meme warfare. Then, somewhere in 2014, something happened when viral pranks went toxic, and Gamergate hit its ultimate low point—from which perspective media scholars are totally behind the curve.[16]

A possible answer to this question might lie in the refusal to deal with memes as isolated digital objects that can be reassembled randomly. As discussed previously, memes arise from and are shaped by their progenitor cultural soup, the stock of which is narrative. As Nick Srnicek told me:

> We need new stories, and that's different from just thinking about counter-memes or stopping the flow of information. It's a different temporality effectively, but a new narrative then provides the basis for more immediate responses via social media, memes, etc. There is a narrative to Trump and the rising far right, for instance. And it's a seductive narrative for some people, which then gets expressed in various forms. The left is, mostly, missing that narrative. We need to get to the heart of the matter, rather than attempting to deal with symptoms.
>
> How, then, do we create this story? A number of obstacles stand in our way.

Narrative meme creation is not easily prescriptive; the Clinton campaign, if you remember, could not construct and manage its own memes. Nor is this tight scripting necessarily desirable, as a program of memetic engineering comes with its own authoritarian undertones. But it's equally clear that we can't work with the narrative we have now; we are clearly losing. We are stuck between a rock, a hard place, and another rock; we must move forward.

A good starting point may be to understand just how the right got its narrative in the memetic canon. As Johannes from the Viennese art collective Monochrom observes: "You need a lot of user/follower/

creator-power to really create outreach. 4chan only became the breeding ground of super-memes because of their sheer endless pool of Darwinian non-archival users, some of them online for almost the entire day—and that for years." Matt Goerzen backs this up:

The alt-right memes are so successful due to their bottom-up, populist nature. I've come to understand image board memes as a toolset that can be put to different uses, but only where they fit the job at hand. Memes can be effectively weaponized, as in shitposting on Twitter, a form of cognitive denial of service attack, to use Rand Waltzman's term.

Waltzman, for the uninitiated, is best known around the meme community for spearheading the US Department of Defense Advanced Research Projects Agency's (DoD ARPA) Social Media in Strategic Communications program, which sought to understand how content and ideas spread online and how it might be utilized for military purposes. The combat motif pervading "the meme wars" is not far from an accurate characterization; the left is indeed losing ground because of them. According to free software thinker and Anonymous historian Gabriella Coleman, we simply cannot afford to forego memes:

When the alt-right was gaining ground and various journalists were horrified that images and emotions could "tug" at people and sway them politically, I was equally horrified that they were so naive and negative about emotions and visual culture. Yes, progressives and leftists must include memes and humor in their arsenal to fight back at some quarters of the right and to steer some portion of the internet-crazed youth toward the left. Without it, we will lose a huge base of people. Whether this can be designed through a group effort or must bubble up from below is a whole other question. My sense is that it would be more effective coming from a subcultural base rather than an elite art vanguard.

For the right, 4chan serves the purpose of the subcultural base. It was not that they memed away until they had an album of good images that carried their political narrative, it was that there was no way for their community to have any kind of narrative or shared identity without memes. Being that everyone is anonymous and there are no static users to grant any kind of continuity between posts, the only way for anything to gain any sort of traction on the platform is for it to be replicated ad infinitum. Together, these tiny shards of content come together to form an ideology. So, as Jodi Dean says in *Crowds and Party*, it is up to us to go "beyond the fragments"—to come up with a narrative that, like the alt-right's, can communicate large amounts of meaning via bits and pieces of content. Dean:

> It will be a good experiment to see if meme wars can be effective in undermining the right (that is, making them appear unappealing and undesirable to potential supporters). The challenge is creating bubble-breaking memes since most memes tend to circulate within bubbles of people who already agree. But even if your memes don't break bubbles, they can still be effective if they inspire the left. Bernie Sanders' Dank Meme Stash was a fantastic source of fun and inspiration during the US election.

Is there a surefire way to bring the next Dank Meme Stash into being? The answer seems doubtful. Johannes reminds us that creating political memes is a PR approach to internet culture.

> People sniff out PR very fast. And in the end it can turn against you and your campaign. I understand the need to create easily shareable counter-info-memes, but that's pretty much already happening. A ton of good images are already circulating in the specific bubbles. But how to get out of the bubble? You can't penetrate conservative bubbles with liberal content. Your content has to be so obscure and mysterious that it's not working as a propaganda tool anymore. Or will just be used for ridicule.

The level at which this ambiguity can be created and still hold its political force, however, is extremely high. Goetzen believes that there is an effective way to weaponize memes for ideological purposes, by steering ones which are already popular and meaningful for a contested demographic.

> This aligns with the "redirect method" that attempts to counter violent extremism circles. The idea of designing or topdowning memes (or "forcememing" in the parlance of imageboard culture) is a pretty

challenging task. Many of the government types I've spoken with in elucidating these questions over the past months have ideas about how this can be done, but it involves pretty vast resources, and more resembles the sort of work done by Cambridge Analytica than anyone in the image-board or alt-right cultural orbit.[17]

(Ironically, Cambridge Analytica's algorithm has its roots in memetic science; a significant part of Christopher Wylie's research before he was brought in was identifying how Crocs had become so popular.)

So we don't have the resources to "forcememe", and we can't create a meme stash without being inauthentic, and we can't lose the meme wars, so what can we do? While leaving this section with the same dilemma we started with, at least now we have a productive foundation from which to approach our final instructive example.

MEMEFEST'S COMMUNICATION DESIGN AND MEMETIC TEMPORALITY

Memefest (Subvert. Create. Enjoy.) is an annual design and advertising competition that defines itself as an "alternative community for the discussion of radical and critical communication projects." Founded in 2002 and originally based in Slovenia, Memefest describes itself as a "festival of radical communication, which encourages students, academics, artists, professionals and activists to interrogate the commercialization of everyday life, focusing on the media and visual communication environment."[18] The P2P Foundation wiki has an interesting page on the original idea and early history of Memefest.[19] The aim was to do something against "the dominant ideologies of modern consumer culture [that] are inherently toxic to our physical and mental environment."

Participants engaged in an "educational process through their creative engagement with a specific theme and through written feedback that a list of the 60 best preselected submissions received from all members of the jury." Memefest positioned itself explicitly as a critique of the design and advertising world. "Most design events are grounded on spectacle and a decontextualized approach to design, while serving as one of the primary mechanisms that define what is good work and who's an excellent author/designer/communicator."

I consulted Oliver Vodeb, member of the Memefest Kolektiv and founder, editor and curator of the Memefest Festival of Socially

Responsive Communication and Art. At some point Vodeb moved to Melbourne and the festival moved with him. In his new role as overworked academic teacher, I asked Oliver about his original ideas:

> I heard about memes in the late nineties as an undergraduate student, in a time when the concept became more known within media activist circles. I remember reading Rushkoff's Media Virus, which I really liked, and then I engaged with more academic theories of memetics. The explosion of internet-driven media activism and networking at around 2000, together with a deep enthusiasm about new media's liberating potential, made memes an attractive concept.[20]

Back then, Oliver explains, "one would see a memetic principle in action on a regular basis and because back then the Internet was by far not as regulated and centralized as it is today, people would be able to spread memes in ways that seemed really promising." Memes had so much potential because of their open character, a notable distinction that also keeps Oliver away from conceiving of them as programmable and deployable like others have thought.

> They were living and independent entities that want to infect a potential carrier. As a concept they provided a semiotic and rhetoric distance, which was useful as it implied that the quality of memes in terms of their social impact is something that is not necessarily given, but rather develops in time through selection.

Time, it seems, turns out to be the lynchpin. Give memes 15 years to develop since memefest, and now they're in bed with the alt-right.

> I agree with Bernard Stiegler when he speaks about the destruction of attention and the resulting destruction of care. Technology enables this every day. Memes seem to like an environment dominated by any kind of drugs. Unfortunately, bad memes result largely from a culture dominated by social networks like Facebook, produced and consumed by people on bad drugs.

Instead, Oliver suggests that the left should think first and foremost about pleasure and the pharmacological aspects of media, design and communication—and only later about the specifics of content and arguments. While memes can build stories, today's data/media sphere is largely losing its narrative. The destruction of the narrative corresponds with what Rushkoff calls "the constant now": "In order to replicate the feeling of pleasure that we gain from the states of being in the constant now, we need to be fed bits of media without a narrative context because the instant, temporal gratification is what brings pleasure."

MEMES AS DIALECTIC IMAGES

In a three part essay series written in early 2018, Marc Tuters and I have tried to bring critical meme theory forward. Starting off from a McLuhan angle, we describe memes as "cultural byproducts of the app ecosystem; the medium, not the meme, is the message. Memes are eyewash of an optimization arms race that strives to reach as far down into the limbic system as possible." Yet, this media theory approach ignores the political question about what's to be done with a primarily right-wing meme culture that is now dominating the internet channels (and thus the minds of its young audiences). The essay triptych proposes reading this transgressive meme culture alongside the Frankfurt School framework in order to find out what is (and isn't) useful from this rich theoretical toolbox that seems to resonate so closely with today's condition. "In the alt-right's memes that frequently pictures Trump as a Teutonic warrior king, Adorno would have diagnosed the symptoms of this very psycho-social situation of male existential resentment concerning loss of power which is considered as the conditions of possibility giving rise to fascism."[21] Such a dark judgment can be contrasted with the populist vitalism of Adorno's colleague Walter Benjamin, who saw popular culture as subversive in contrast to the fundamentally conservative quality of elite culture.

In the second contribution we applied Benjamin's ambiguous term "dialectic image" to memes, a reference to his idea that we need a new visual epistemology comparable to cinematic montage. Originating from the 1999 film *The Matrix*, the red pill/blue pill meme can be interpreted as a kind of vernacular gloss on the Benjaminian idea of awakening. "You take the blue pill—the story ends, you wake up in your bed and believe whatever you want to believe." The red pill, in contrast, awakens one to the brutal truth of modernity, in relation to which Benjamin once famously quipped: "There is no document of civilization which is not at the same time a document of barbarism."[22] One need barely ask how

Benjamin would react to the film's Platonic allegory—of course he would take the red pill. Yet somewhat unfortunately, the red pill has become a code word for neo-reactionary enlightenment, a symbol of awakening from the false consciousness of liberal political correctness.[23]

The third article opens with an analysis of the Harambe gorilla, a meme centering around a sacrificial animal that captured the imagination of the internet in the Brexit/Trump summer of 2016. Not only did the alt-right manage to instill Harambe with a greater purpose, but it was the meme's very polyvocality that made it so politically effective—an organizational capacity that we should not underestimate. As an open-ended symbol, a kind of totem for condensing the disparate set of grievances of an insurgent neo-reactionary form of identity politics, the alt-right's use of Harambe conformed to a classical trope in fascist discourse, what we called the "theft of joy" narrative, in which an authoritarian figure channels the desires and resentments of the "radical loser" through a spectacle of collective hate.[24]

To investigate North America's alt-right is one thing. To start building a non-fascist, post-biological foundation for progressive image circulation and radical "memelabs" is another. This is where European initiatives could step in and pick up on the earlier histories presented here. Such experiments could dream up speculative concepts and apply these to software and image prototypes. Instead of docile metaphors such as replication and virality, the starting point could be the fabrication of singular data filled with strange beauty leading us out of the defensive mode of identity-based political correctness. Instead of fear, let there be freedom. Instead of resentment, we reclaim liberation. We want to build a counter hegemony based on imagination, not on "rights". This will be (for once), not a movement of lawyers. We need to get out of the formalistic, defensive mode. Official reality is not a castle worth defending.

Another step to get there would be to re-imagine forms of organization that appeal to this age. We may have to get rid of the ostensibly neutral scientific reference of the lab in the first place, discarding terms such as memelab. Imaginative labels matter. But what's most important is not to walk away from the issue. Social media have become an integral part of our political theatre. Its logics will only eat deeper into the system. To believe that we deal here with fads or fashions that will soon blow over is dangerously naïve. This is why we have to work on memes as myths and not simply brush it aside. Even if we want to overcome memes and not utilize this particular form of image/text distribution, we will have to go this way. The answer to the alt-right will ultimately require rebuilding social media architecture from scratch, deconstructing the platform idea itself, and creating a scalable successor to our distributed network logic. Apart from the need for narratives and related visual culture, there's an even stronger demand to address the issue of acceleration. Should alternative memes circulate at the same speed as the overall internet (or even faster, as the avant-garde dreams of)? Are we running out of time? What about slow memes? What if the real-time regime itself turns out the core of the problem? According to Franco Berardi, we need an alternative rhythm of elaboration—to slow down our blurred sequentiality, to heal from relentless acceleration, and to instead find a new rhythm, a new movement. This cannot be accomplished through further acceleration. Real-time communication already ruins our bodies and our minds. A collective move back to the drawing board will be worthwhile. According to Berardi, the digital realm is leading to "decorpetization", a movement toward a "bodiless brain."[25] The infosphere is one giant nervous stimulation. Before we can even start telling the new narrative, what we need is a "reconfiguration of mental elaboration."

10

Before Building the Avant-Garde of the Commons

"Power is invisible, until you provoke it." GFK[1]—"Bread and circuses for everyone, *wealthfare* for the elites and welfare for the restless disenfranchised." ZeroHedge—"'Rock stars' are arrogant narcissists. Plumbers keep us all from getting cholera. Build functional infrastructure. Be a plumber." Molly Sauter—"We live in the golden age of ignoring smart people." Zak Smith—"The multitude which is not reduced to unity is confusion." Pascal—"We lost the fight for the Internet. But the battle against central authority remains." Peter Sunde—"We may be decentralized and disagree on a lot of topics amongst ourselves, but operations are always carefully coordinated." Anonymous

According to Albert Camus, we are living in nihilism… We shall not get out of it by pretending to ignore the evil of our time or by deciding to deny it. The only hope is to name it, on the contrary, and to inventory it to discover the cure for the disease… Let us thus recognize that this is a time for hope, even if it is a difficult hope…[2]

To take up Camus' challenge, I will discuss at least one form in which alternatives present themselves—not via tactical media, social media alternatives or organized networks but through the broader concept of the commons, one ideally placed to form a bridge between tech issues and society at large. I am not selling hope as a fake solution here. As Slavoj Žižek notices about his *Courage of Hopelessness*, this is a dark book as well. Žižek prefers to be a pessimist: "not expecting anything, I am here and there nicely surprised, while optimists see their hopes dashed and end up depressed all the time."[3]

In *Out of the Wreckage*, British activist-writer George Monbiot called us to go beyond critique and instead understand the power of the simple narratives so often used by the powers that be. "Our minds appear to be attuned to particular stories that follow consistent patterns." We need a compelling narrative. "Without a coherent and stabilizing narrative, the movements remain reactive, disaggregated and precarious, always at risk of burnout and disillusion."[4] Monbiot's advice: "The only thing that can replace a story is a story." Let's turn the commons into one of those stories. Before we condense the somewhat boring and flat legal commons concept and turn it into an attractive meme we can all embrace, let's see where we stand in the commons debate.

French philosopher Frédéric Neyrat asked: "How is it possible to imagine a different future, an alternative future, without impeding it with our imaginations? Nowadays we live in anticipation societies which, paradoxically aim to ward off the future."[5] At the end of *Futurability*, Franco Berardi sees the forces of darkness "trying to subdue thought, imagination and knowledge under the rule of greed and the rule of war."

Against evil he holds the space of possibility, "stored in the cooperation among the knowledge workers of the world." He's certain that a "social brain" must exist. "As long as we are able to imagine and to invent, as long as we are capable of thought independent from power, we will not be defeated."[6] Do we have the *potency* to do this? Berardi defines potency as "the energy that transforms the possibilities into actualities."[7] In his 2013 dialogue with Berardi, Mark Fisher remarked: "We need to reclaim the future, and that means recovering a *prospective* time, where we are not endlessly protesting against or obstructing capital, but thinking ahead of it. Here is the space for art to reinvent itself—as the site for a multiplicity of visions of a post-capitalist future." Alternatives will have to include elements of pop culture, carnival, a public sphere of the imagination and play, not the boredom and seriousness of the Habermasian coffee shop, says Henry Jenkins.

Operating inside the contexts of technology, media activism and internet politics, the commons is finally turning into a hotly debated topic outside of theory and activist circles. Code is shaping our world, and its architecture is voluntary and plastic. However, written by geeks and engineers, this code is anything but God-given, let alone neutral. Where do the underlying ideas come from and how are we going to accelerate the transition? Who's taking the lead?

What seems highly conceptual and speculative on one day, locks in millions a few months later. Everyone who witnessed the late 1990s schism between free software and open source will know what's at stake with a concept such as the commons. Will we have a reformist pro-business commons next to a marginal radical and politically correct one? Ideas matter—and no more so than in the case of the commons. Discussion outcomes matter. And if the crucial commons are increasingly technological, who's in charge of the law, when "code is law"? Will the debate on the nature and architecture of commons in the end get into the hands of lawyers? How do we turn the commons into a lively and diverse political strategy that brings people together in order to reinvent public infrastructure?

This chapter addresses two issues: the search to come up with a workable definition of the commons, and the question of who's going to design and prototype it. I propose a reinvention of the (artistic) avant-garde notion as "organized networks", a concept I have worked on over the past decade with my Sydney friend, the media theorist Ned Rossiter.[8] The argument here is a call to move away from idealistic notions of "what we have in common" toward a materialistic understanding of actually existing commons as both small-scale experiments and large-scale infrastructures. This is matched with a twenty-first-century organizational model that is capable of inventing the future, creating workable concepts (aka running code) within a post-capitalist framework that are strong enough to obstruct the inevitable—the all-too-predictable, all-too-depressing appropriation machine. Whereas most artists, activists, designers and researchers have so far focused on laboratory scale in(ter)ventions, discussions in various contexts show that it is now time to scale-up and remove the neo-liberal privatization dominance over infrastructure.

FROM COMMON(S) TO INFRASTRUCTURE

Let's work our way through the multitude of terms and definitions, from Hardt and Negri's commonwealth to the commons, community and communism. I was never attracted to any of these. I have always preferred working within smaller social units, from friendships and groups to networks and movements, Not the Big We but the small we. I am neither a liberal who believes in copyright reform, nor a communist who sticks to the Gosplan. As children of the Age of Difference, my generation grew up in the shadow of disastrous communes. Overseen by gurus under the close guidance of this religion or that ideology, they advocated a Total Sharing Experience, from joints and food to partners and income. Emerging from this wreckage, it was no surprise that—despite the drawbacks of the "tyranny of informality"—I preferred the openness of networks and movements over the closed totality of The Group and related "folk politics". I never read my own refusal or inability to scale up as a personal tragedy or trauma. Instead, I was—and remain—a strong believer in a diverse ecology of interconnected, autonomous DIY infrastructures that function as a blueprint for larger public initiatives in the near future. Such "islands in the net", to use Bruce Sterling's phrase, shift their roles depending on the local political conditions—sometimes functioning as future labs, at other times merely operating as defensive shields that preserve subversive practices.

What we have in common cannot be discussed without the element of liberation—and (individual) liberty. Freedom means being liberated from the limiting social norms of the "communalis". Such a definition of freedom is, in the conservative politically correct context of the early twenty-first century, often discredited as individualist and capitalist. However, from an activist perspective, this is not at all the issue. Liberation from the tribe of the family, the mob of the village, or the gang of the factory, paves the way for future experiments with yet unknown social formations such as the "free association of peers". How do we build committed long-term relationships that don't slip into the ruts of boredom and routine? How do we create a culture that is open to change outside of the conventional legal frameworks?

Over the past decade, a group of small scale "minoritarian" practices have emerged, largely ignoring the theoretical debates about communism in the society at-large and instead creating actually-existing commons. Think free software, Wikipedia and Creative Commons (the alternative copyright license, mostly used for music and publications). Or consider all the initiatives that the Peer-to-Peer Foundation lists on its impressive web resource.[9] Yet Creative Commons is a reformist approach inside intellectual property law—and thus a domain of lawyers. As Gary Hall observes in his *Pirate Philosophy*: "Exponents of this understanding of copyright have been able to form a 'coalition of experts with the legal access and resources' to mount a powerful campaign that frequently over-shadows often more interesting and radical approaches."[10] Admittedly "copyleft"[11] goes further than Creative Commons.[12] However it is still a legal contract. Ultimately, then, it forces its legal will upon others via the Power of the Law, threatening any potential violators with repressive sanctions.

Ever since the rise of neo-liberalism and the decline of the welfare state, the construction of infrastructure can no longer be taken for granted. This has led to a dual (if not schizo) approach of the commons, which splits into two distinctive directions. On the one hand there is the grassroots, bottom-up approach, in which the commons is seen by both state and market actors as a productive—or even utopian—concept in the making. On the other hand, there is the top-down imaginary, in which the commons contributes to a renaissance of state-owned public infrastructures.

Over the years I've become a fan of the concept of the minimal commons, a set of implicit social practices and agreements that is so invisible, informal and direct, that it no longer needs lawyers and contracts (not even smart ones), a form of the social contract (meta-smart or uber-smart!) that drifts into our habitual realm and then segments into the collective unconscious. As a lived reality, it feels self-evident. This wouldn't mean that the element of trust has to be eliminated (a techno-libertarian proposal I never understood, let alone supported). Direct does not mean that we have nothing in common with the rest. It means instead a liberation from the repressive, inward-looking aspect of the constructive "community" constantly in need of reaffirmation.

Ideally the commons is an open infrastructure that frees us up precisely because it can be assumed. It should be designed to be taken for granted (at least from moment to moment). Good infrastructure is self-evidently enjoyed, not noticed. It is simply there, and works. We should not always have to "work on the commons".

Common infrastructure aims for radical openness; it's not "our" commons, against "community standards" that lead to the "ignore/delete" logic, applied by the moderation factories (as documented in *The Cleaners*[13]). The commons should distance itself from the "identitarian" politics that so easily turns into a repressive force to keep the group, movement or party together. We versus They. Us against the "Other". In contrast, good infrastructure is accessible to the public, placed in common hands. Once privatized, it quickly deteriorates, becoming (too) expensive and may no longer be maintained.

The commons-as-concept is described in somewhat similar terms by Lauren Berlant. Her question is what a commons means in times when things fall apart or break down. According to Berlant, "the commons is an action concept that acknowledges a broken world and the survival ethics of a transformational infrastructure. This involves using the spaces of alterity within ambivalence."[14] Commons are only beginnings.

"Through the commons the very concept of the public is being reinvented now, against, with, and from within the nation and capital."

For Berlant, the commons is not some utopia. Instead, "it points to what threatens to be unbearable not only in political and economic terms but in the scenes of mistrust that proceed with or without the heuristic of trust."[15]

My ideal commons is not just self-evident infrastructure. At times it can also be a place of lively debate and disagreement. It is not a place of consensus. The commons I have in mind consists of dozens of factions. It is a place where people gather and discuss, as they did during the recent occupations of public squares and universities across the globe. As Roberto Esposito writes, quoted in Gary Hall: "The Commons is a place where the interests of a large number of diverse groups... come together but also exist in a state of tension and conflict and are in fact often demonstrably incompatible and incommensurable."[16] It is this aesthetic meta-structure that we can call the commons. It is both metaphysical (in terms of the law) and material.

WHO WILL BUILD THE COMMONS?

Ever since the dark 1970s we've been hearing that the avant-garde is over, dead, history. Read as many art history catalogues as you like; you'll never bring it back to life. The avant-garde was part of a historical narrative. But that chapter has been closed, its ideas and approaches long buried. "The avant-garde is dead" is the art equivalent of Thatcher's "There is no alternative." Mediation and art that transcends our current confines is no longer possible. We're stuck in the virtual cage, forever. There is no authentic communication possible anymore. There's a beauty of the fall, the moment you lose yourself, but even that dissipates. Every activity has already been retweeted; every event incorporated into your Facebook News Feed before the situation has fully unfolded. There is no original time/space experience possible of "speaking with" before the representation of "speaking to" sets in.

In Peter Bürger's *Theory of the Avant-Garde* (1973) the question of organization does not come up. By then avant-garde thinking had turned into the history of things long gone. Bürger's generation had turned into academic outsiders. The very notion of avant-garde had been ceded, turned into the exclusive domain of literary scholars and art dealers. From then on, the avant-garde was synonymous with modernism and the experts were eager to reduce the subsequent styles and schools to stylistic techniques such as collage and montage. Theorists and critics internalized their role as aesthetic observers, stuck summarizing the pre-war debates between Adorno and Lukács and relating them back to Kant, Schiller and Hegel. An entire generation was socialized to study their own society through the real mirror of the nineteenth century, with major roles given to Marx and Nietzsche. Bürger is a prime example of this trend. The idea of having this avant-garde theory engage with a group like the Situationists (who dissolved in 1972), conceptual art, technological experiments or minimal music was simply out of the question. The sarcasm, cynicism and despair of the post-war years, the depression amidst the rise of totalitarian regimes was a primal energy that theory wasn't able to catch.

In the aftermath of the roaring 1960s, the historical chain of the avant-garde came to an end. The continual lineage (albeit turbulent and contested) of different schools, movements and groups that gathered and debated, wrote manifestos and developed a common visual grammar, had been disrupted for good. No one has yet been able to fix this. There were plenty of experiments but most of them drifted toward pop culture with the aim to diffuse, to slow down, to relax. The ascetic or even militant aspect of the avant-garde no longer held any appeal.

The Situationists, partisans of the supersession of art, were acutely aware of the fact that they were last descendants of the "historical avant-garde". The group explicitly played with the unavoidable desire to be forgotten. What counted was a radical negation of the presence and the abolition of memory and melancholy. Demoralize your fans and your friends, dissolve into nothing, withdraw to the zero position. In his book on Guy Debord, *Revolution in the Service of Poetry*, Vincent Kaufmann declares that a situationist who reveals himself is suspect. "To truly be a situationist, one must forget situationism in general, and Debord in particular, whose desire for obscurity was fulfilled. Real revolutionaries know how to make themselves forgotten, disappear, lose themselves. Their fame resides only in their vocation for obscurity, the standard against which their subversive potential must be measured."[17] All these insights were impossible to learn in an art school or university seminar. Insurrection only rises from the shadows, a condition that first needs to be created. Obscurity is the a-priori, the starting point of every heresy. Become the first and last guardian at once; organize the lived moment.

These ideas had to be lived in the here and now—and then promptly forgotten.

McKenzie Wark's 2011 chronicle of the Situationist International, *The Beach Beneath the Street*, denied that the group was ever an artistic avant-garde in the first place. The succession had fallen apart.[18] From now on, the aim was to move beyond art. And art practice could only be overcome, in a Hegelian sense, through a "brutal evolution". The Situationist International (SI) as the first post–avant-garde movement began to position art as merely one of many creative practices. The objective was to establish a multi-disciplinary diversity inside the group, an attack strategy aimed directly at the painters and their traditional exhibition strategies. Strictly following postmodern instructions, the group sought to exercise the Baudrillardian aesthetics of disappearance. As Guy Debord once said, "the SI knew how to fight its own glory."[19] Art could only be tolerated if it undermined the unique gesture.

In our distant understanding of the "historical" avant-garde, movements like the Situationist International were still membership organizations, cliques of friends that hang out in the same cafes and openings. Their leader, Guy Debord, ran the network as if it were a Trotskyist sect. The stories of expulsions within the group are numerous—and notorious. They strongly suggest that there must have been something at stake. Wark echoes this idea, pointing at the death cult energy that came with the repeated excommunication of SI members. "Exclusion of living members meant social death." The SI "wrestled with the problem of how to make collective belonging meaningful, as something requiring some sacrifice. The possibility of exclusion made participation in the Situationist game meaningful."[20] As groups and movement post-SI had no members and no membership administration in the first place, expulsion as a formal act, disappeared.

Forty years later, Saskia Sassen gave expulsions another meaning and context.[21] Since the 2008 global financial crisis, expulsions no longer refer to the correct set of beliefs, but to the banking practice of disowning and evicting house owners who can no longer pay their mortgage. These days, it rarely happens that a member is officially removed from a group. We're addressed as users, supporters, volunteers, followers, at best as temporary employees—but never as members. The same can be said of those who are fired from their jobs. Nowadays, one's contract simply expires (in the same way as the rent contract terminates and is no longer renewed). People aren't fired, they "lose" their jobs in a fatal manner (in the same way as they found it, those lucky bastards).

For tomorrow's avant-garde, the question of how (not) to deal with membership and how to design internal commitment becomes strategic. How can we overcome the user status? In today's social media society, weak ties seem to be symbolic of all social relationships. We have yet to design how strong ties operate, or define an alternative to the strong-weak binary. This is the domain of the organized networks, a concept that has been around since 2005 and whose time will come. While these "strong tie" experiments are only just emerging, we will see a natural erosion of Facebook's weak ties as the dominance of the intrusive social media mega-platforms draws to a close.

The Situationists soon disappeared into oblivion, dissipating into a cloud of cigarette smoke and alcohol. They were to be replaced by postmodernism, an ahistorical condition that proclaimed diversity and fragmentation, in which, by definition, an avant-garde position was no longer possible. Any singular vector dissolved into a hodgepodge of pointers and loops. Amidst all the quotations and pastiches, what would following a leading aesthetic school mean anyway? Wisely, conceptual art no longer presented itself as avant-garde; it refrained from making claims outside of the art system itself.

Who will organize the visual arts? It is not going to be the art market, with its dealers, collectors and gallerists. Nor will it be the curator class, with their fly-by-night projects and their global biennales. In the past this task was undertaken by networks of artists, magazines and journals. Today, many look at websites like *e-flux* and *Hyperallergic* to take up this task, in the same way as *Art Forum* and *Texte zur Kunst* were influential in the 1990s. Editorial decisions steer the global conversation. Or at least that's the premise. Yet one of the problems here is the decline in influence of (print) journals, zines, pamphlets and text in general. Dominated by academics, theory and criticism are such niche activities that they are no longer capable of mobilizing any organizational capacity outside of their own small (yet global) circles. Others, like artist and activist Greg Sholette, co-editor of *Collectivism after Modernism* from 2007,[22] have explored the organizational potential of art collectives. Can networks take up this role, and if so, what architectures should they have?

The question of organization cannot merely be discussed under the rubric of the institution-as-such. This would inevitably lead us into a dead-end street. Institutions can only control discourse, they are incapable of producing new styles and trends (let alone novel internet memes). Organization as rebirthed institution would be nothing more than bureaucracy eating its own children. A twenty-first-century avant-garde is neither working for the Party, nor for the Institution (called contemporary arts) but situates itself within the web of infrastructures necessary to secure collective and individual freedom. Avant-garde movements have never existed long enough to become institutions. In fact, today's #1 paranoia, to become institutionalized, was never a problem in the past. Left to art historians, galleriots and cultural policy makers, collective units are split up into individual life stories that can better be marketed and sold. Mind you, there is no situationist museum—and not even one for surrealism. It would deserve to be burned down in the first place. In the past, the remnants of avant-garde groups have protested such historical repackaging, attempting to stop the administration and commercialization of the past.

These days the challenge is to overcome the perpetual present. How can there be a dialectics in the ever present now? Being a forerunner is a project with a clear expiry date. How can a group or network achieve today's mission to "destroy worlds", as Andrew Culp phrased it?[23] How can we de-familiarize ourselves with social media, detaching ourselves from its grasp? We need to escape the cage and start again on a journey. This is one of the strong original internet myths: surfing. This type of info détournement is analogous to an alcoholic or psychedelic derive. Web surfing may not be toxic but it certainly feels like a psychic journey...

We're spreading a dangerous message here. Today, organization is perceived as synonymous with terrorist cells, conspiracy theories, organized crime and secret societies on the one hand, and bureaucratic structures such as NGOs and political parties on the other. Fundamentally organization means coordinating the self as a social entity with other selves. It means going out into the world and acting together—and in a post 9/11 world, this is no longer an innocent move. Organization puts officials in the highest stage of alert, ready to utilize violence. As many have experienced in our *Minority Report* age, this ain't no joke.

This is why terrorists can no longer create cells and gather; the algorithmic prediction machine will immediately spot them. Hiding after the fact is no longer possible; your location is already known. Before the Act, there is no possibility of testing or trialing, no ability to rehearse or conduct a dry run. At best Takfiri terrorists remain silent and invisible, staying under the radar until they strike. After the Act, it is over for them. Every hit is a suicide attack, committed by lone wolves or tiny, isolated groups. There is zero time for organic growth. The lack of trial and error is compensated by an indirect transmission of experiences via mainstream broadcast media. Networks may or may not exist. What does exist are shared experiences, a collective awareness with common references, YouTube videos, links on social media filled with body language and slogans, in short: memes. What counts is the impact of violent memes.

Organization as a concept has been all too quickly co-opted into the corporation. Business Dictionary defines organization as "a social unit of people that is structured and managed to meet a need or to pursue collective goals. All organizations have a management structure that determines relationships."[24] Organization studies became a servant of the academic managerial class (and their bean counters). This has locked away vital knowledge. These days we cannot think of organization outside of the business and management context. There must be someone behind it, steering the wheel. The social (whoever that may be) cannot organize itself. There are only professional structures with an identifiable leadership structure.

So what's on offer besides the business cliché of "standing on the shoulders of giants"? In *Inventing the Future*, authors Smicek and Williams[25] demand the founding of a think-tank. Others have argued for a return to the Party. The Democracy in Europe movement (DiEM25) blends both approaches. Launched early in 2016 by Yanis Varoufakis, DiEM25 mixes a Brussels-based lobby group, a trans-local grassroots movement and a networked think-tank.[26] Going beyond the twentieth century we need to do trial-and-error experiments with contemporary forms of organization that do work.

We also need to find out if there's any future in the avant-garde mode. The question that keeps returning is how the social can take command in the age of social media? Can this only be done from inside our already existing social networking regimes or is there the possibility of an outside position, in which small groups catalyze an exodus out of these walled gardens? "What the Situationists were struggling to achieve was a new kind of collective being, unlike both the Communists and previous avant-gardes such as the Letterists," McKenzie Wark remarks. Is a Third Situationist International still possible in this age of accumulating urgencies, from right-wing populism and platform capitalism?

Wark seems to suggest that Debord did everyone a favor by polarizing the question of creativity, "by choosing paths, rather than allowing the movement to sink, like so many others, beneath the weight of its incoherence."[27] What do such experiments look like 50 years later? What does collective being entail today, if we step beyond the dominant (and highly limited) libertarian premise of "collective self-interest"?[28] In an age where membership has become a technical matter of filling out a CAPTCHA to prove you're not a bot, how does recruitment work? Can we still plot in secret?

According to the academic consensus, the avant garde is an integral part of modernism and thus a thing of the past. As modernism, defined as a historical period, modernism is long gone. We can be nostalgic about the fabulous lives of its icons and eccentrics, but we cannot bring it back to life. All we can do is quote their artistic legacies and visit their retrospectives, dreaming of some increasingly less likely encounter that might radically shake up our everyday life. This is the historical post-modernist condition, a period we (re)entered most recently in 2008, when the aggregation of global crises hit the surface and made an abrupt end to the joyous quotation fest. This crisis is also a crisis of organiza-tion. We cannot run from this issue so easily, expecting that political parties, NGOs and Facebook are sufficient. They are not. We need artistic counter models to the start-up, non-terrorist insurgency models, twenty-first-century prototypes of the "open conspiracy". Bouncing off ideas against the avant-garde approach is merely one of many ways to invent new forms of organization that match our current *zeitgeist*.

The argument here is that we need to see the avant-garde as a social organization, disconnecting it from the question of beauty and modernity and its shock of the new. We no longer have scores to settle with conventional notions such as linear, chronological time, in which the avant-garde projects itself into an imaginary future. In a world dominated by the permanent present, it is the real-time regimes that we need to confront. What is a real-time avant-garde? Is it possible in the first place to bring players together and act in such a short time frame? Can we escape the permanent now in the first place? That's the "present shock," described by Douglas Rushkoff: "If the end of the twentieth century can be characterized by futurism, the twenty-first can be defined by presentism."[29] How can we escape the time trap?

If we stick to the art perspective, the challenges are radically different from a century ago. The task is, no longer to make "anti-art" with the aim of upsetting the Western bourgeois class. Autonomy is today's problem and solution at the same time, creating a whirlpool of opposing expecta-tions in which pop culture and aesthetic singularity have to be achieved simultaneously. All works have to contain multiple layers of interpretation, which nevertheless have to become easily digestible by gallery owners, marketing experts, art critics and the audience. This makes it hard to restage the demand for self-criticism. There are already enough threads and stories, comments and trolls.

Online heresy is the new normal. Art no longer bears "the unique stamp of Greek art," as Peter Szondi once stated.[30] We live in a post-deconstructivist period, tired because we're wired. Everything is already a montage, with endless layers of data, software, content, form and meaning stacked on top of each other. A century ago the "destruction of coherence" was experienced as a shock. Today it is the new normal. Instead of adding another layer or creating another image, our avant-garde will fight on the invisible and immaterial front lines, from the shadows, as invisible networks, without links or Likes or recommendations, working on "data prevention". As Debord insisted, tomorrow's revolutionaries should practice some intentional amnesia, "not bringing us up as a reference, forgetting us a little."[31] That's the crystal of today's act of organization.

Notes

INTRODUCTION

1. Twitter, July 11, 2017.
2. Laura Penny, "Who Does She Think She Is?" https://longreads. com/2018/03/28/who-does-she-think-she-is/.
3. Mara Einstein: "If a friend tells us they liked the latest *Jurassic Park* movie, there's no reason for us not to believe it. Unfortunately, what we also come to believe is that amassing friends on Facebook or followers on Twitter is ultimately about sharing with compatriots. It is not: it is about creating an audience for advertisers. Our relationships, then, become means of facilitating market transactions, or in the parlance of the market; they have been monetized." (*Black Ops Marketing*, OR Books, New York, 2016, p. 8).
4. http://highscalability.com/blog/2018/8/22/what-do-you-believe-now-that-you-didnt-five-years-ago-centra.html.
5. Siva Vaidhyanathan, *Anti-Social Media*, Oxford University Press, New York, 2018, p. 10.
6. In his article, "Debunking the Biggest Myths About 'Technology Addiction'" (https://undark.org/article/technology addiction-myths/), Christopher Ferguson claims that, contrary to other research that spread "moral panic", technology is not a drug, not a mental illness and does not lead to suicide. These are stats wars amongst psychologists that are caught in the biases of their own empirical reality, produced by their research parameters. My point here is to be wary of the medicalization of everyday language.
7. Slavoj Žižek, *The Year of Living Dangerously*, London, Verso, 2012, p. 127.
8. Society of the Social is not just a fun reference to Guy Debord's *Society of the Spectacle*, but a provocation to the virtual absence of traditional sociology in the social media debate. The concept can also be read as an extension of an earlier essay dating back to 2012 called "What is the Social in Social Media?" (published in *Social Media Aby*ss, Polity, Cambridge, 2016).
9. Claude Levi-Strauss, *Tristes Tropique*, Penguin Classics, London, 2011, p. 408.
10. Previous volumes: *Dark Fiber*, MIT Press, 2002; *My First Recession*, V2/NAi, 2003; *Zero Comments*, Routledge, 2007; *Networks without a Cause*, Polity, 2012; *Social Media Abyss*, Polity, 2016.
11. Clifford Geertz, *The Interpretation of Cultures*, Basic Books, 1973, p. 15.
12. Franco Berardi, *Futurability, The Age of Impotence and the Horizon of Possibility*, Verso, London/New York, 2017, p. 172.
13. Bernard Stiegler, *Technics and Time, 2, Disorientation*, Stanford University Press, Stanford, 2009, p. 3.

14. The Invisible Committee, *Now*, Semiotext(e), South Pasadena, 2017, p. 48.
15. Ulises Mejias, *Off the Network, Disrupting the Digital World*, University of Minnesota Press, Minneapolis, 2013. I have also used Pepita Hesselberth's overview "Discourses on dysconnectivity and the right to disconnect", in New Media & Society, 2018, Vol. 20(5), pp. 1994–2010.
16. For an overview, see Pepita Hesselberth, "Discourses on dysconnectivity and the right to disconnect," in: *New Media & Society*, 2018, Vol. 20(50), pp. 1994–2019.
17. https://thedisconnect.co/. "We believe you should be able to disconnect from the internet without sacrificing the possibilities of a digital platform. By forcing you to physically disable your internet connection, *The Disconnect* creates a dynamic that allows you to enjoy engaging digital content at your own pace."
18. In 2011 Nathan Jurgenson argued that we should abandon the digital dualist assumption that the on and offline are separate. "Social media has everything to do with the physical world and our offline lives are increasingly influenced by social media, even when logged off. We need to shed the digital dualist bias because our Facebook pages are indeed "real life" and our offline existence is increasingly virtual." https://thesocietypages.org/cyborgology/2011/09/13/digital-dualism-and-the-fallacy-of-web-objectivity/.
19. Andrew Keen, *How to Fix the Future*, Atlantic Books, London, 2018, p. 192.
20. Keen, p. 41.
21. Written in response to, and inspired by Bernard Stiegler, *Automatic Society*, Volume 1: The Future of Work, Polity Press, Cambridge, 2016, pp. 6–18.
22. Dieter Bohn, "Google's Most Ambitious Update in Years," *The Verge*, May 8, 2018, www.theverge.com/2018/5/8/17327302/android-p-update-new-features-changes-video-google-io-2018. Thanks to Michael Dieter for contributing to this research.
23. See: www.androidauthority.com/youtube-take-a-break-864783/.
24. Slogan of the www.wellbeing.google.com website.
25. Simone Stolzoff, "Technology's 'Time Well Spent' movement has lost its meaning," https://qz.com/1347231/technologys-time-well-spent-movement-has-lost-its-meaning/.
26. The Invisible Committee, *Now*, Semiotext(e), South Pasadena, 2017, p. 16.
27. Maja van der Velden, "Another design is possible—Looking for ethical agency in global information and communication technology," PhD thesis, University of Bergen, 2009 (www.globalagenda.org/).
28. Mark Fisher, "Optimism of the Act," www.k-punk.org/optimism-of-the-act.
29. "The growth of right-wing forces is ominous," interview with Noam Chomsky, June 22, 2018, www.frontline.in/politics/the-growth-of-rightwing-forces-is-ominous/article10108703.ece.

1. OVERCOMING THE DISILLUSIONED INTERNET

1. An earlier draft of this chapter was published in June 2017 in *E-flux Journal* #83: www.flux.com/journal/83/141287/overcoming-internet-disillusionment-

on-the-principles-of meme-design/. The meme design part was moved to Chapter 8.

2. Designer-researcher Silvio Lorusso (http://silviolorusso.com/), who provided valuable comments on this piece, noted that a similar relativism has taken over visual culture. This might be the reason why trained, professional graphic designers are the least equipped to produce effective memes. In response, memes are often associated with underground amateur culture. Meme creation is therefore often described as a mysterious process, for instance in the documentary on "the first meme," the Techno Viking (as discussed in Chapter 9).

3. Slavoj Žižek on Trump, Brexit, and fake news, *Channel 4 News*, February 13, 2017. www.youtube.com/watch?v=ByKXcIPi7MI.

4. See: www.computing.co.uk/ctg/news/3036546/decentralising-the-web-why-is-it-so-hard-to-achieve.

5. Jarett Kobek, *I Hate the Internet*, Serpent's Tail, London, 2016, p. 25.

6. Danah Boyd, "Did Media Literacy Backfire?" January 12, 2017, http://dmlcentral.net/media-literacy-backfire/.

7. In 2018 OR Books published a reprint of Dorfman and Mattelaert's 1971 classic *How to Read Donald Duck*: https://en.wikipedia.org/wiki/How_to_Read_Donald_Duck.

8. Cathy O'Neil, *Weapons of Math Destruction*, Penguin, London, 2016.

9. See: www.cjr.org/tow_center/donald_trump_media_organization.php. "In many ways, Donald Trump sees himself not just in opposition to the existing press but in competition with them, too." The piece argues that through Twitter and various other channels, Trump is running his own media organization (thanks to Marc Tuters for the reference).

10. Dan P. McAdams, "The Mind of Donald Trump," *The Atlantic* (June 2016), www.theatlantic.com/magazine/archive/2016/06/the-mind-of-donald-trump/480771/.

11. Ibid.

12. Tara Burton, "Apocalypse Whatever," *Real Life* (December 13, 2016), http://reallifemag.com/apocalypse-whatever/.

13. Materials on the proposed "organized networks" concept that I have written about over the years together with Ned Rossiter have been brought together in *Organization after Social Media*, Minor Compositions, 2018. Download the book for free here: www.minorcompositions.info/wp-content/uploads/2018/06/organizationaftersocialmedia-web.pdf.

14. Here it could also be relevant to mention the New York psycho-historian Lloyd de Mause (https://en.wikipedia.org/wiki/Lloyd_deMause), whose 1984 study *Reagan's America* could be read as a source of inspiration for us today.

2. SOCIAL MEDIA AS IDEOLOGY

1. An earlier version of this chapter appeared in *e-flux Journal #75*, September 2016: www.e-flux.com/journal/75/67166/on-the-social-media-ideology/.

2. www.netimperative.com/2016/04/facebooks-context-collapse-massive-drop-personal-sharing/. There is growing evidence that first-hand personal experiences are no longer shared by Joe and Joanna Sixpack. Nicholas Carr calls this "context restoration". "When people start backing away from broadcasting intimate details about themselves, it's a sign that they're looking to reestablish some boundaries in their social lives, to mend the walls that social media has broken (..) They are shifting their role from that of actor to that of producer or publisher or aggregator." www.roughtype. com, April 10, 2016.

3. Sherry Turkle, *Reclaiming Conversation*, Penguin Press, New York, 2015.

4. Charles Leadbeater. *We-Think*, Profile Books, London, 2008, p.1.

5. When terms or symptoms are inflated, they lose their meaning. This might also be the case with addiction. If entire societies are addicted, the term loses its ability to create differences and it is time to search for alternative concepts. One possible concept could be the term "stickyness". Julia Roberts on social media: "It's kind of like cotton candy: It looks so appealing, and you just can't resist getting in there, and then you just end up with sticky fingers, and it lasted an instant."

6. See the announcement for the launch of the Unlike Us network, July 2011: http://networkcultures.org/unlikeus/about/.

7. Variation of the popular drawing of Freud with a naked lady worked into his forehead entitled "What's On a Man's Mind", a poster that also decorated my teenager bedroom in 1976–7.

8. Reference to Louis Althusser's famous essay, "Ideology and the Ideological State Apparatuses (Research Notes)", which first appeared in La *Pensée 151*, June 1970.

9. All quotes are from Wendy Chun, "On Software, or the Persistence of Visual Knowledge," *Grey Room 18*, MIT Press (Cambridge), (Winter 2004), pp. 26–51.

10. Michel Houellebecq, *Whatever*, Serpent's Tail, London, 1998, p. 29.

11. Term introduced by James Wallman in his book *Stuffocation: Living More With Less*, Penguin Books, London, 2015.

12. Term taken from a presentation by Anne Helmond and Carolin Gerlitz at the *Unlike Us #2* symposium in Amsterdam, March 8–10, 2012: http://networkcultures.org/unlikeus/2012/03/10/anne-helmond-and-carolin-gerlitz-explain-the-like-economy/.

3. DISTRACTION AND ITS DISCONTENTS

1. First published in German (translated by Andreas Kallfelz), in *Lettre International* 120, 2018. A shorter version in English was published by Eurozine: www.eurozine.com/distraction-and-its-discontents/. The full version, edited by Ed Graham, can be found here: http://networkcultures. org/geert/2018/03/27/distraction/.

2. www.axios.com/sean-parker-unloads-on-facebook-2508036343.html.

3. www.theguardian.com/technology/2017/oct/05/smartphone-addiction-silicon-valley-dystopia.
4. www.theverge.com/2017/12/11/16761016/former-facebook-exec-ripping-apart-society.
5. Siva Vaidhyanathan, *Anti-Social Media*, Oxford University Press, New York, 2018, p. 35.
6. Heinrich Geiselberger, *The Great Regression*, Polity Press, Cambridge, 2017 (original in German, published by Suhrkamp Verlag, 2017). The term democracy fatigue is the title of the opening essay by Arjun Appadurai. The term *great regression* refers to the growing income disparity in the West since the 1980s (https://en.wikipedia.org/wiki/Great_Regression). The anthology was an initiative by Suhrkamp editor Heinrich Geiselberg, intending to come up with an international response to Donald Trump, Brexit and the rise of right-wing populism in Europe. The book was translated into 14 languages (www.thegreatregression.eu/).
7. Eva Illouz, *Why Love Hurts*, Polity Press, Cambridge, 2012, p. 90.
8. www.news.com.au/technology/online/social/leaked-document-reveals-facebook-conducted-research-to-target-emotionally-vulnerable-and-insecure-youth/news-story/d256f850be6b1c8a21aec6e32dae16fd
9. Hans Schnitzler. *Kleine filosofie van de digitale onthouding*, De Bezige Bij, Amsterdam, 2017.
10. Title of a "technology fair" in Seoul: www.artsonje.org/en/abetterversion ofyou/.
11. An example from the arts world would be the London exhibit at Furtherfield, "We're All Addicts Now". www.furtherfield.org/events/are-we-all-addicts-now/.
12. www.reddit.com/r/Showerthoughts/comments/7dki8w/surfing_the_web_has_become_like_watching_tv_back/
13. I am using interpassivity here in a media technological sense, slightly different from Robert Pfaller who defined the term as "delegated enjoyment" and "flight from pleasure". In my definition, interpassivity expresses the dialectical move backwards toward "e-regression", after the visual turn of the web. In chronological terms, the algo-recommendation mode of viewing, browsing and swiping, encapsulated inside the social media ecology, can be seen as successor of the post-war passive consumption of limited channels, followed by engaged watching during the age of cable and satellite TV (as theorized by cultural studies) and the interactivity empowerment of the user during the 1990s (the new media age). See: Robert Pfaller, "On the Pleasure Principle in Culture," *Illusions Without Owners*, Verso Books, London, 2014, p. 18.
14. https://jameshfisher.com/2017/11/08/i-hate-telephones.
15. Sherry Turkle, *Reclaiming Conversation, The Power of Talk in a Digital Age*, Penguin Press, London/New York, 2015.
16. http://knowyourmeme.com/memes/distracted-boyfriend.
17. See also the related topic of "read receipts": www.dailydot.com/irl/swipe-this-read-receipts/.

18. Roland Barthes, *A Lover's Discourse: Fragments*, Penguin, London, 1990, p. 87.
19. https://en.wikipedia.org/wiki/Daydream.
20. Freud calls the function of "dreamwork" secondary revision, providing the motives with a narrative cohesion. The dream gets a comprehensible meaning. See Sigmund Freud, *The Interpretation of Dreams*, Barnes and Noble Books, New York, 1994, p. 155.
21. Wendy Chun, *Updating to Remain the Same*, The MIT Press, Cambridge (Mass.), 2017, p. 1.
22. Ibid, p. 6.
23. Ibid, p. 12.
24. Mark Greif, *Against Everything, On Dishonest Times*, Verso Books, London/ New York, 2016, p. 225.
25. www.socialcooling.com/.
26. The Data Prevention Manifesto by the Plumbing Birds: http://dataprevention. net/.
27. Petra Löffler, *Verteilte Aufmerksamkeit, Eine Mediengeschichte der Zerstreuung*, Diaphanes, Zürich-Berlin, 2014. A summary of the book and interview with the author I made, in English, can be found here: https:// necsus-ejms.org/the-aesthetics-of-dispersed-attention-an-interview-with-german-media-theorist-petra-loffler/.
28. The subtitle of Bernard Stiegler, *Technics and Time, 2*, is Disorientation, Stanford University Press, Stanford, 2009. Here Stiegler points at the double movement at play here: disorientation both as a loss of direction, going all over the place, and as a loss of the "Orient", the imaginary place of otherness (p. 65).
29. Rob Horning: "Distraction is no longer a relief from tedium but its metronome." https://thenewinquiry.com/blog/ordinary-boredom/.
30. Roland Barthes, *Camera Lucida*, Hill and Wang, New York, 1982, p. 43.
31. www.forharriet.com/2017/09/self-care-after-incense-burns-out.html.
32. Email exchange, December 12, 2017.
33. As Gerald Moore put it: "The same drug that, when consumed in a toxic environment, further mires us in toxicity, can in different circumstances enable us to project visions for environmental transformation. And it follows that the key for therapy, surely, is to build *pharmaka* that facilitate, rather than inhibit, the construction of alternatives." Quoted from: http:// pharmakon.fr/wordpress/on-the-pharmacology-of-the-dopamine-system-fetish-and-sacrifice-in-an-%e2%80%98addictogenic-society%e2%80%99-gerald-moore/.
34. See interview with Pepita Hesselberth: http://blogs.cim.warwick.ac.uk/ outofdata/2017/06/01/on-disconnection/.
35. The term "vital information" has been defined in the context of the Amsterdam Seropositive Ball (1990) and was further elaborated by City of Amsterdam Chief Science Officer, Caroline Nevejan: "By 'vital' I mean information that supports an individual in his or her specific circumstance. (..) For information to be vital, it has to touch upon our natural presence

physically or socially. Mediated presence, which generates vital information, will also ultimately have this effect, (..) it is information that matters from the perspective of the receiver." Caroline Nevejan, *Presence and the Design of Trust* (University of Amsterdam PhD), 2007, pp. 174–6. www.being-here. net/page/375/vital-information.

36. City of Amsterdam Chief Science Officer Caroline Nevejan wrote to me: "The other day I was not told a person in our neighborhood had passed away. It had been announced on Facebook, which I do not check, so I missed the funeral. I would have wanted to know, someone could have phoned me. Yet, when confronted with the choice to engage with Facebook and its siblings, I made the choice to not engage and accept that I will miss things. If ever I need the social networks to get my vital information, I will engage. The longer I can postpone the better; I accept the collateral damage that is caused by my non-participation." (email, December 11, 2017).

4. SAD BY DESIGN

1. "Having my phone closer to me while I'm sleeping gives me comfort." Quote from research by Jean M. Twenge, "Have Smartphones Destroyed a Generation?" *The Atlantic*, September 2017, www.theatlantic.com/magazine/ archive/2017/09/has-the-smartphone-destroyed-a-generation/534198/. Twenge observes that "teens who spend more time than average on screen activities are more likely to be unhappy." She sees a decrease in social skills. "As teens spend less time with their friends face-to-face, they have fewer opportunities to practice them. In the next decade, we may see more adults who know just the right emoji for a situation, but not the right facial expression."

2. Quotes from Jaron Lanier, *Ten Arguments For Deleting Your Social Media Accounts Right Now*, Argument 7: Social Media is Making You Unhappy, Vintage, London, 2018, pp. 81–92.

3. Adam Greenfield, *Radical Technologies*, Verso Books, London, 2017, p. (thanks to Miriam Rasch for the reference).

4. Contrast this with the statement of Amos Oz: "You probably recall the famous statement at the beginning of Anna Karenina, in which Tolstoy declares from on high that all happy families resemble one another while unhappy families are all unhappy in their own way. With all due respect to Tolstoy, I'm telling you that the opposite is true: unhappy people are mainly in conventional suffering, living out in sterile routine one of five or six threadbare clichés of misery." (*The Black Box*, Vintage, 1993, p. 94, thanks to Franco Berardi.)

5. Earlier I dealt with the psychopathology of information overload, in part influenced by the writings of Howard Rheingold, for instance in my 2011 book *Networks Without a Cause*. While this diagnosis may still be relevant, psychological conditions such as sadness come in when we're online 24/7, the distinction between psyche and phone has all but collapsed and we're no

longer administrating incoming information flows on large screens in front of us via dashboards.

6. I am using the exhaustion here in the way Gilles Deleuze once described it, in contrast with feeling tired. Unlike tiredness we cannot easily recover from exhaustion. There is no "healing tiredness" (Byung-Chul Han) at play here. Take Teju Cole's description of life in Lagos: "There is a disconnect between the wealth of stories available here and the rarity of creative refuge. Writing is difficult, reading impossible. People are so exhausted after all the hassle of a normal Lagos day that, for the vast majority, mindless entertainment is preferable to any other kind. The ten-minute journeys that take forty-five. By day's end, the mind is worn, the body ragged." (*Everyday is for the Thief*, Random House, New York, 2014, p. 68).

7. In his blog post *Social Media as Masochism*, Rob Horning writes: "Much of social media is a calculated effort to 'accumulate' esteem and grant agency. Self-consciousness of ongoing social media use could trigger an intense need to escape from self. Social media, he proposes, "has affordances to make 'self-construction' masochistic and self-negating." "One puts an aspect of oneself out there to dream of it being mocked, and that pain of mockery disassociates us from the deeper vulnerabilities of the 'real self' that is being deferred and protected for the moment." https://thenewinquiry.com/blog/social-media-as-masochism.

8. Byung-Chul Han, discussing Carl Schmitt's friend-enemy distinction in the Facebook age, in: *Müdigkeitsgesellschaft*, Matthes & Seitz, Berlin, 2016, p. 71.

9. Adrienne Matei, "Seeing is Believing, What's so Bad about Buying Followers?" http://reallifemag.com/seeing-is-believing/. The essay deals with the stylized performance of authenticity of Instagram images and their accompanying captions that "often denounce superficiality and strategic image manipulation and emphasize the value of embracing rather than concealing imperfections." She observes that for influencers, "authenticity tends to be bound up with aspiration: an image is 'true' if it captures and triggers desire, even if the image is carefully and even deceptively constructed. The feeling it inspires in the midst of scrolling is what matters." Using something faked, edited, misleading, or out of context to attract attention isn't the platform's problem but its point. Matei argues that, while there may not be fake images, there are fake audiences. As one influencer explained: "It's not so much outrage as people pity you. It's like people who pay for all the drinks at the bar just to feel like they have friends. It's sad." Matei concludes: "Buying followers can alleviate hassle, but it entails embracing the paradox of all counterfeiting: coveting a currency whose legitimacy you are in the process of undermining."

10. This is written with William Styron's *Depression* in mind, Vintage Minis, 2017, written in 1990, in respect of all those that suffer from severe forms of depression.

11. Quotes from Mark Greif, *Against Everything*, p. 225.

12. Ibid.

13. Ibid., p. 227.

14. Quotes from interview with Audrey Wollen by Tracy Watson in *Dazed*, November 23, 2015, www.dazeddigital.com/photography/article/28463/1/girls-are-finding-empowerment-through-internet-sadness. See also www.instagram.com/audreywollen/ and www.tumblr.com/search/audrey%20wollen (thanks to Miriam Rasch).
15. Interview with Audrey Wollen by Yasi Salek for *Cultist Zine*, June 19, 2014, www.cultistzine.com/2014/06/19/cult-talk-audrey-wollen-on-sad-girl-theory/.
16. Vladimir Nabokov described toska as "a sensation of great spiritual anguish, often without any specific cause. At less morbid levels it is a dull ache of the soul, a longing with nothing to long for, a sick pining, a vague restlessness, mental throes, yearning. In particular cases it may be the desire for somebody of something specific, nostalgia, love-sickness. At the lowest level it grades into ennui, boredom." More here: https://advokatdyavola.wordpress.com/2012/05/07/an-elegy-for-passion/ (thanks to Ellen Rutten for the reference).
17. www.captionstatus.com/sad-whatsapp-status/.
18. Julia Alexander, YouTube's top creators are burning out and breaking down en masse, June 1, 2018, www.polygon.com/2018/6/1/17413542/burnout-mental-health-awareness-youtube-elle-mills-el-rubius-bobby-burns-pewdiepie.
19. www.bustle.com/articles/162803-what-is-a-snapchat-streak-heres-everything-you-need-to-know-about-snapstreaks.
20. Taylor Lorenz, Teens explain the world of Snapchat's addictive streaks, where friendships live or die, April 14, 2017, www.businessinsider.com/teens-explain-snapchat-streaks-why-theyre-so-addictive-and-important-to-friendships-2017-4.
21. www.ingeniotechsarl.com/how-to-receive-your-boyfriends-whatsapp-messages-online-using-android-phones-without-jailbreak-2018.
22. The two blue check marks appear when all participants in the group have read your message. Alternatively, you can long press on a message to access a "message info" screen, detailing the times when the message was received, read or played. Users can check "last seen" indicators on the top left of a conversation to know when a contact was last in the app, but the blue check marks are more direct. One can disable the feature, though WhatsApp will "punish" you by not letting you see what others have "last seen" online. There is contractual power here in who sets the rules. It is not reciprocity; it is a mutual obligation, both toward the app and to the contacts.
23. www.quora.com/Why-do-I-feel-anxiety-while-using-WhatsApp#
24. https://vulcanpost.com/71431/why-im-not-keen-on-whatsapp-blue-ticks/.
25. Suzannah Weiss, "Why We Swipe Right And Then Ignore Our Tinder Matches," May 10, 2016, www.bustle.com/articles/157940-why-we-swipe-right-and-then-ignore-our-tinder-matches.
26. Email to the author, August 7, 2018.
27. Susan Sontag, *Under the Sign of Saturn*, Vintage Books, New York, 1981, p. 111.

28. Ibid. Sontag writes: "Slowness is one characteristic of the melancholic temperament. Blending is another, from noticing too many possibilities, from noticing one's lack of practical sense."

29. See analysis of Aristotle in Raymond Klibansky, Erwin Panofsky, Fritz Saxl, *Saturn and Melancholy*, Basic Books, New York, 1964. I have used the updated German translation, *Saturn und Melancholie*, Suhrkamp, Frankfurt am Main, 1990, pp. 76–92.

30. Blue is not only the color of Facebook, Twitter and IBM, there's an avalanche of medical stories on popular news websites about the dark side of blue light in terms of sleep deprivation. See for instance: www.health.harvard.edu/ staying-healthy/blue-light-has-a-dark-side.

31. What does this impossibility to access melancholy mean for the imagination, if we stick to Julia Kristeva who once asserted that "there is no writing that symbolically refers to love, and no imagination that is not openly and secretly melancholy." See: Julia Kristeva, "On the Melancholic Imaginary," *New Formation* number 2, (Fall 1987). It is Marc Fisher who does have a kind of melancholia he can relate to, one which "consists not in giving up on desire but in refusing to yield. It consists in a refusal to adjust to what current conditions call 'reality'—even if the cost of that refusal is that you feel like an outcast in your own time…" (*Ghosts of my Life*, Zero Books, Winchester, 2014, p. 24).

32. In that sense, sadness is an unexpected side effect of the social media business. According to Wolf Lepenies, both historical sketches of utopian societies and the twentieth-century avant-garde promised to overcome the hesitation to act that came with this bourgeois disease. A true revolutionary is not melancholic. In some instances melancholy was even forbidden (*Melancholie und Gesellschaft*, Suhrkamp, Frankfurt am Main, 1998, p. 40).

33. Whitney Joiner, "Why is Melissa Broder So Sad Today?" *Elle Online*, March 14, 2016.

34. Laura Bennett in *Slate Magazine*, quoted in *Elle Online*, March 14, 2016.

35. All quotes from the chapter "Help Me Not Be a Human Being," in: Melissa Broder, *So Sad Today*, Scribe, London, 2016, pp. 29–36.

36. Selected tweets from Melissa Broder's @sosadtoday twitter account, May–July 2018.

37. Thanks to Katharina Teichgräber for giving me the reference.

38. Slavoj Žižek, *The Courage of Hopelessness*, Penguin, London, 2017, p. xi. He writes: "We have to gather the strength to fully assume the hopelessness."

39. Paul B. Preciado, *Testojunkie*, Feminist Press, New York, 2013, p. 304.

40. Translation of a phrase from Hans Demeyer, "Uitgeput op drift, Over Mark Fisher," *De Witte Raaf* 193, (mei-juni 2018). p. 5.

41. http://networkcultures.org/entreprecariat/the-designer-without-qualities/.

42. "Western Melancholy/ How to Imagine Different Futures in the Real World," http://interakcije.net/en/2018/08/27/western-melancholy-how-to-imagine-different-futures-in-the-real-world/#rf3-2509.

43. See interview with George Didi-Huberman in *Liberation*, September 1, 2016, Tears are a Manifestation of Political Power, www.liberation.fr/

debats/2016/09/01/georges-didi-huberman-les-larmes-sont-une-manifestation-de-la-puissance-politique_1476324 (thanks to Marie Lechner for the reference).

5. MEDIA NETWORK PLATFORM: THREE ARCHITECTURES

1. Quote from the 1971 Dutch television debate between Noam Chomsky and Michel Foucault, www.youtube.com/watch?v=3wfNl2LoGf8. Transcript here: https://chomsky.info/1971xxxx/.
2. This chapter can be read as an update of earlier remarks in Geert Lovink, *Social Media Abyss*, Polity Press, 2016, pp. 2–3. An early German version, based on a lecture in Salzburg was published as Medien-Netze-Plattformen, in *MedienJournal*, 41. Jahrgang, nr. 1/2017, p. 26–32. No doubt there's a biographical element in the choice and following order of the three terms. I started off as an organizer of alternative media, becoming a media theorist who got into computer networks, advocating "net criticism" and then having to confront myself, and the Institute of Network Cultures that I started in 2004, with the might of centralized platforms.
3. See https://internetworldstats.com/stats.htm.
4. See Geert Lovink & Pit Schultz, *Jugendjahre der Netzkritik*, Institute of Network Cultures, Amsterdam, 2010.
5. Florian Cramer points at the use of another term: "In the last decade, German humanities have developed a broad, general and trans-historical notion of media as 'mediality' ('Medialität') in which any material or imaginary carrier of information qualifies as a medium, from CPUs to angels. The notion of 'medium' has thus replaced and superseded the older semiotic-structuralist term of the 'sign'." https://web.archive.org/web/20090420094737/http://medienumbrueche.uni-siegen.de/groups/medienwissenschaften/blog/. He adds: "Perhaps, looking back, you could say that 'German media theory' is something like 'krautrock': a phenomenon that was seen and constructed abroad but was much less understood as a cohesive phenomenon inside Germany." (email, September 3, 2018). See also Claudia Breger, Zur Debatte um den "Sonderweg deutsche Medienwissenschaft", Zeitschrift für Medienwissenschaft, Zürich, 1/2009, pp. 124–7.
6. See https://en.wikipedia.org/wiki/Network_science and http://network sciencebook.com/.
7. Zeynep Tufekci, *Twitter and Tear Gas, The Power and Fragility of Networked Protest*, Yale University Press, London, 2017, p. 76.
8. See my interview with Sebastian Gießmann on his book *Die Verbundenheit der Dinge*, in Necsus, Spring 2016, https://necsus-ejms.org/philosophy-weaving-web-interview-media-theorist-sebastian-giessmann/.
9. One of many entry points could be Orit Halpern's description of the turn from language to landscape in the work of cybernetics pioneer Kepes. Along these lines platforms could be defined as a "landscape of forms". Halpern describes Kepes' mid-century "reconfiguration of cognition, perception and sense to algorithm, pattern, and progress..." Kepes saw the constant data and

interface interactions as an environment, a paradigmatic shift that led, for instance to Kevin Lynch's 1953 study *The Image of the City*, written under Kepes' supervision. See Orit Halpern, *Beautiful Data*, Duke University Press, Durham, 2014, pp. 80–81.

10. Paul Langley & Andrew Leyshon, Platform Capitalism: The Intermediation and Capitalization of Digital Economic Circulation, in: Finance and Society, 2016.

11. Nick Srnicek, *Platform Capitalism*, Polity Press, Cambridge, 2016, p. 43.

12. Andrew McAfee & Erik Brynjolfsson, *Machine Platform Crowd*, Norton & Company, New York, 2017, p. 163.

13. *Machine Platform Crowd*, p. 164.

14. Interestingly, Bratton teaches at the same institute at the University of California in San Diego as Manovich once used to work, before he returned to New York in 2013.

15. Benjamin Bratton, *The Stack*, MIT Press, Cambridge (Mass.), 2016, p. 5.

16. Ibid., p. 37.

17. Ibid., p. 64.

18. Ibid., p. 41.

19. Ibid., p. 52.

20. Bernard Stiegler, *The Neganthropocene*, Open Humanities Press, London, 2018, pp. 127–38. A first draft of this chapter was presented there.

21. *Platform Capitalism*, p. 128.

22. Arguably, the concept "public stack" was for the first time discussed during a day-long boat tour called the *Public Stack Summit* in the Amsterdam vicinity on June 19, 2018, organized by Waag Society as part of the We Make The City festival. The announcement reads: "For the past 25 years, we've allowed the market to govern the development of the internet. Now we're reaping what we've sown. Data centralisation, surveillance capitalism and the 'commercialisation of every click', are all part of our online reality. We need to set the stage to design, build and foster the open, fair and inclusive internet we deserve."

23. This thesis has been developed together with Pit Schultz in Berlin, July 2018.

24. Theodor W. Adorno, *Minima Moralia*, Suhrkamp Verlag, Frankfurt am Main, 1951, p. 57.

25. Reference to Bernard Stiegler, *The Lost Spirit of Capitalism*, Polity Press, Cambridge, 2014.

6. FROM REGISTRATION TO EXTERMINATION, ON TECHNOLOGICAL VIOLENCE

1. https://news.vice.com/en_us/article/8xbbqg/facebook-hired-an-outside-group-to-investigate-its-role-in-myanmar-ethnic-cleansing.

2. Jonathan Beller, *The Message is Murder*, Pluto Press, London, 2018, p. 2.

3. All quotes from Yuval Noah Harari, *Homo Deus*, Vintage, London, 2017, pp. 408–11.

4. See https://en.wikipedia.org/wiki/Gerrit_van_der_Veen. A Wikipedia page on the 1943 attack only exists in Dutch and German.

5. Detlef Hartmann, *Die Alternative: Leben als Sabotage*, Zur Krise der technologischen Gewalt, iva-Verlag, Tübingen, 1981, p. 4.

6. This could be an ideal place where the critical discourses of Simondon and Stiegler on technology and individuation can be inserted. I won't do that here but the link is obvious.

7. In 1978 I took an SPSS course on how to use punch cards that were fed into an IBM mainframe computer, a compulsive part of my BA in political science at the University of Amsterdam.

8. For instance in danah boyd and Kate Crawford, "Critical Questions for Big Data: Provocations for a Cultural, Technological, and Scholarly Phenomenon," *Information, Communication, & Society*, 15:5 2012, pp. 662–79.

9. Edwin Black, p. 7.

10. Ibid., p. 8.

11. Ibid., p. 208.

12. Ibid., p. 209.

13. Ibid., p. 343.

14. Ibid., p. 322.

15. Ibid., p. 104.

16. Ibid., p. 425.

17. Ibid., p. 424.

18. Ibid., p. 306.

19. Ibid., p. 313.

20. See https://en.wikipedia.org/wiki/Database.

21. Kenneth Werbin, *The List Serves*, Population Control and Power, Institute of Network Cultures, Amsterdam, 2017.

22. Andrew Culp, *Dark Deleuze*, University of Minnesota Press, Minneapolis, 2016, p. 19.

23. Culp, p. 2.

24. Instead of merely "working on the myth" in the tradition of Hans Blumenberg (*Arbeit am Mythos*), as Thomas Rid suggests in his introduction (p. xviii), we should disrupt the narrow cyber myth itself and broaden our scope, even beyond the obvious anti-militarist, gender and post-colonial approaches.

25. This part is based on an earlier blog posting on the *4th conference of the Communist Alliance Ums Ganze on the Question of Technology* in Hamburg, November 24–26, 2016. https://techno.umsganze.org/en/programm/.

26. Website of Capulcu: https://capulcu.blackblogs.org. The publication was connected with an event (30.9–2.10.2016) in Cologne entitled "Big Data >Esc >Del >Shutdown >Reboot: Life is no Algorithm—Collective Perspectives against the Technological Attack." (https://bigdata.blackblogs.org).

27. Capulcu redaktionskollektiv, *Disrupt!* Widerstand gegen den technologischen Angriff, Unrast Verlag, Münster, 2017.

28. Michael Hardt and Antonio Negri, *Assembly*, Oxford University Press, Oxford, 2017, p. 119.
29. Assembly, p. 110.
30. Assembly, p. 111.

7. NARCISSUS CONFIRMED: TECHNOLOGIES OF THE MINIMAL SELFIE

1. An earlier version of this chapter appeared in *The Failed Individual: amid exclusion, resistance, and the pleasure of non-conformity*, edited by Katharina Motyl and Regina Schober, Campus Verlag, Frankfurt, 2017.
2. Bruno Bettelheim, *The Informed Heart*, Avon Books, New York, 1960, p. 260.
3. http://blogs.getty.edu/iris/whats-the-difference-between-a-selfie-and-a-self-portrait/.
4. Alicia Eler, Mirrors Multiply the Selfie: The Doppelgänger Dilemma, July 8, 2013, http://hyperallergic.com/74877/mirrors-multiply-the-selfie-the-doppelganger-dilemma/.
5. Taken from YouTube recording of lecture *Why I love Selfies and Why you should too (damn it)* by Katie Warfield, Vancouver, March 26, 2014.
6. Peter Bürger, *Theory of the Avant-garde*, University of Minnesota Press, Minneapolis, 1984, p. 48.
7. Christopher Lasch, *The Culture of Narcissism*, Warner Books, New York, 1979, p. 22.
8. Christopher Lasch, *The Minimal Self*, W.W. Norton & Company, New York, 1984. p. 16.
9. See, for instance, the project of the St. Petersburg founders of the facial recognition app Findface, the photographers Artem Kukharenko and Alexander Kabakov. With Findface one can compare a picture taken in public space with the databases of the selfies online, for instance of the Russian social media site VKontakte. The software is extremely precise and makes it possible to find out a person's identity in an anonymous crowd with a success rate of 70%.
10. Frequent references to the individuation concept can be found in Bernard Stiegler, *Uncontrollable Societies of Disaffected Individuals*, Polity Press, Cambridge, 2013, pp. 101–25.
11. More on this in Bernard Stiegler, *The Lost Spirit of Capitalism*, Cambridge, 2014, pp. 72–4.
12. Michel Foucault, About the Beginning of the Hermeneutics of the Self: Two Lectures at Dartmouth, in *Political Theory*, 21:2, May 1993, pp. 198–227
13. Sherry Turkle, *Reclaiming Conversation, The Power of Talk in a Digital Age*, Penguin Press, New York, 2015.
14. Jodi Dean, Images without Viewers: Selfie Communism, 1. February 2016, http://blog.fotomuseum.ch/2016/02/iii-images-without-viewers-selfie-communism/.

15. Adilkno, *Media Archive*, Autonomedia, Brooklyn, 1998, p. 12–15.
16. Quotes from email interview with Alex Foti, June 15, 2016.
17. Quotes from email interview with Ana Peraica, June 20, 2016.
18. See http://networkcultures.org/online-self/ for conference program, blog reports. videos of the lectures, interviews and related postings.

8. MASK DESIGN: AESTHETICS OF THE FACELESS

1. www.ctrl-verlust net/glossar/kontrollverlust/.
2. This chapter consists of material that has been co-written over the years together with Daniel de Zeeuw (University of Amsterdam) and Patricia de Vries (Institute of Network Cultures/Erasmus University), who have both written their PhDs on related topics. I am very grateful for their collaboration and extensive comments on this book version that I have compiled from different sources (written in Dutch and English) and have rewritten for this purpose.
3. See Daniel de Zeeuw, Immunity from the image: The right to privacy as an antidote to anonymous modernity, in: *Ephemera*, volume 17(2), p. 259–81. www.ephemerajournal.org/contribution/immunity-image-right-privacy-antidote-anonymous-modernity.
4. Bogomir Doringer: "The unstable identity of the present begs for the return of the power of the mask from ancient times, when it was used as a form of protection, disguise, performance, or just plain entertainment." (http://facelessexhibition.net/statement) References may vary from the classic Venetian carnival masks and masquerade ball masks to skeleton makeup for Halloween, the Palestinian keffiyeh worn during demonstrations, or the rows of women dressed in scarlet cloaks, with oversize white bonnets obscuring their faces, based on the red-and-white costume from Margaret Atwood's *The Handmaid's Tale*. www.theguardian.com/world/2018/aug/03/how-the-handmaids-tale-dressed-protests-across-the-world.
5. See http://generation-online.org/p/fpcaillois.htm.
6. Claude Lévi-Strauss, *Der Weg der Masken*, Suhrkamp, Frankfurt am Main, 2004, p. 131. We either can visit the Paris Musée du quai Branly to study traditional masks or browse through the collection of the New York fashion designer Abasi Rosborough, aimed against facial recognition. www.highsnobiety.com/p/abasi-rosborough-fw18-in-plain-sight-campaign.
7. Elias Canetti, *Crowd and Power*, Penguin Books, Middlesex, 1973, p. 434.
8. *Crowd and Power*, p. 438.
9. Gabriella Coleman, *Hacker, Hoaxer, Whistleblower, Spy*: The Many Faces of Anonymous, Verso Books, New York, 2014, p. 64.
10. Brian Knappenberger, *We Are Legion*, January 2012. More information on http://en.wikipedia.org/wiki/We_Are_Legion.
11. Robin Celikates and Daniel de Zeeuw, "Botnet Politics, Algorithmic Resistance and Hacking Society," *Hacking Habitat, Art of Control*, Naio10 Publishers, Rotterdam, 2017, p. 217. www.academia.edu/24789171/Botnet_

politics_algorithmic_resistance_and_hacking_society_in_Hacking_
Habitat._Art_of_Control_.

12. Coleman, second edition, p. 416.

13. Coleman, second edition, p. 416.

14. Quoted from selected excerpts of their 1995 book *Mind Invaders*: http://
lutherblissett.net/archive/215_en.html.

15. See for instance this English text: www.republicart.net/disc/artsabotage/
afrikagruppe01_en.htm.

16. See: http://facelessexhibition.net/statement. There's a catalogue available,
edited by Bogomir Doringer and Brigitte Felderer, *FACELESS: Re-inventing
Privacy Through Subversive Media Strategies*, De Gruyter Verlag, 2018.

17. www.mediamatic.net/360812/en/faceless-statement.

18. In her essay on Zach Blas, Hito Steyerl mentions a few more names: "Laura
Poitras, Metahaven, Jesse Darling, Sang Mun, Tyler Coburn, Dmytri Kleiner,
Andrew Norman Wilson and James Bridle, and organizations such as Auto
Italia South East," *ArtReview*, March 2014 FutureGreats issue, http://
artreview.com/features/2014_futuregreats_zach_blas/.

19. See www.facelessexhibition.net/frank-schallmaier.

20. For more information on her work, please visit www.hesterscheurwater.
com/.

21. www.facelessexhibition.net/shahram-entekhabi.

22. Patricia de Vries, https://platformjmc.files.wordpress.com/2017/08/de-
vries-august-2017.pdf

23. This part is taken from the essay by Patricia de Vries and Geert Lovink,
"Against A Calculated Life: How to Overcome the Privacy Worldview" in:
Hou Hanru/Luigia Lonardelli (ed.), *Please Come Back, The World As Prison*,
Mousse Publishing, Rome, 2017, pp. 74–84.

24. https://platformjmc.files.wordpress.com/2017/08/de-vries-august-2017.
pdf.

25. Nicolas Thoburn, To Conquer the Anonymous: Authorship and Myth in the
Wu Ming Foundation, in: *Cultural Critique* 78, Spring 2011, p. 127.

26. http://nettime.org/Lists-Archives/nettime-l-1403/msg00000.html

27. Email exchange with Michael Dieter, February 15, 2017.

28. For more indications in this direction, see the small but very rich anthology
Opaque Presence: Manual of Latent Invisibilities edited by Andreas
Broeckmann and Knowbotic Research, Diaphanes, Zurich, 2010.

29. https://en.wikipedia.org/wiki/Whisper_(app).

9. MEMES AS STRATEGY: EUROPEAN ORIGINS AND DEBATES

1. Thomas Rid, *Rise of the Machines*, Scribe, Brunswick, 2016, p. 48.

2. Quoted from the Wikipedia page on memes: https://en.wikipedia.org/wiki/
Meme.

3. Email correspondence, September 14, 2017. Email in possession of the
authors.

4. W.J.T Mitchell, "The Work of Art in the Age of Biocybernetic Reproduction," *Modernism/modernity* 10, no. 3, September 2003. https://muse.jhu.edu/article/46443/pdf.

5. Morris Kolman, *I Have No Mouth and I Must Meme: Internet Memes, Networked Neoliberalism, and the Image of the Economic*, BA thesis, Williams College, 2018. https://unbound.williams.edu/theses/islandora/object/studenttheses%3A1204.

6. Angela Nagle, *Kill All Normies*, Zero Books, Winchester, 2017, p. 67.

7. Rid, p. 49.

8. Gerfried Stocker & Christine Schöpf (ed.), *Memesis, The Future of Evolution*, Ars Electronica Catalogue, Springer Verlag, Wien, 1996, p. 9.

9. James Gardner, "Memetic Engineering." *Wired*, May 1, 1996 www.wired.com/1996/05/memetic/.

10. Mark Dery, Wild Nature, in: *Memesis, The Future of Evolution*, p. 213.

11. A critique of bio-medical metaphors from 1990s cyberculture till today is a somewhat separate topic that can easily be extended. Take this quote: "Memes are a time-proven way to express disdain, mocking and other sentiments while maintaining an air of light humour. They can therefore be helpful in a transition away from full-throated, deeply-felt outrage. Memes are the methadone of the Internet. They are only to be generated or consumed under advice of a doctor, as prolonged use might result in undesirable side effects." https://medium.com/rally-point-perspectives/the-end-of-memes-or-mcluhan-101-2095ae3cad02.

12. Richard Barbrook, Never Mind the Cyberbollocks..., AEC Forum Memesis, July 25, 1996. http://aec.at/meme/symp/panel/msg00076.html.

13. Vilém Flusser, *Into the Universe of Technical Images*, University of Minnesota Press, Minneapolis, 2011.

14. The original 4:01 Techno Viking video has so far always been available on YouTube, with over 13 million views in 2018: www.youtube.com/watch?v=UjCdB5p2voY. Evidence that Techno Viking is still alive anno 2018 is the fact that the popular video game *Fortnite* added an emote (or dance), called Intensity, inspired by the Techno Viking moves: www.youtube.com/watch?v=xhR0xgWt8bI.

15. Matthias Fritsch, The Story of the Techno Viking (video documentary), 2015, https://vimeo.com/140265561.

16. All interview quotes in this essay come from email exchanges in the February 13–24, 2017 period.

17. Reference to the Cambridge Analytica (UK) big data firm that worked for the Trump campaign. See: https://motherboard.vice.com/en_us/article/how-our-likes-helped-trump-win.

18. See: www.memefest.org/en/.

19. https://wiki.p2pfoundation.net/Memefest. In its first seven years "Memefest established a international network with participants from more than 60 countries and local nodes in Colombia, Australia, Spain and Brazil."

20. http://networkcultures.org/geert/2017/06/22/interview-with-oliver-vodeb-memefest-on-the-addictive-power-of-memes-today/.

21. Geert Lovink and Marc Tuters, *They Say That We Can't Meme: Politics of Idea Compression*, February 11, 2018. https://non.copyriot.com/they-say-we-cant-meme-politics-of-idea-compression/.
22. Walter Benjamin, *Illuminations*. Translated by Harry Zohn. New York: Schocken Books, 2007, p. 256.
23. Geert Lovink and Marc Tuters, *Rude Awakening: Memes as Dialectical Images*, April 3, 2018: https://non.copyriot.com/rude-awakening-memes-as-dialectical-images/.
24. Geert Lovink and Marc Tuters, *Memes and the Reactionary Totemism of the Theft of Joy*, August 12, 2018, https://non.copyriot.com/memes-and-the-reactionary-totemism-of-the-theft-of-joy/.
25. Quotes from Franco Berardi's lecture at the Rietveld Academy, Amsterdam, February 8, 2017, part of the Studium Generale "What Is Happening to Our Brain" lecture series.

10. BEFORE BUILDING THE AVANT-GARDE OF THE COMMONS

1. www.youtube.com/watch?v=gogT3Oog6eg.
2. Back cover text of the NRF Espoir series, written by series director Albert Camus, quoted from Herbert R. Lottman, *Albert Camus, a Biography*, Gingko Press, Corte Madera, 1997, p. 394.
3. Slavoj Žižek, *The Courage of Hopelessness*, Penguin Books, London, 2017, p. xxi.
4. George Monbiot, *Out of the Wreckage*, Verso, London, 2017, p. 6.
5. Quote from Frédéric Neyrat, *Occupying the Future, Time and Politics in the Age of Clairvoyance Societies*, available here: https://atoposophie.files.wordpress.com/2018/05/anticipation-and-politics.pdf.
6. Franco Berardi, *Futurability*, Verso, London, 2017, p. 162.
7. *Futurability*, p. 1.
8. Most recent texts on this topic are Occupy and the Politics of Organized Networks, in: Geert Lovink, *Social Media Abyss*, Polity Press, Cambridge, 2016, pp. 182–204, Geert Lovink and Ned Rossiter, The Politics of Organized Networks, in: Chun & Fisher (ed.), *New Media, Old Media*, Routledge, New York, 2016, pp. 335–45 and Geert Lovink and Ned Rossiter, *Organization after Social Media*, Minor Compositions, Colchester, 2018.
9. See: http://p2pfoundation.net/ and in particular: http://commonstransition. org/. The term "transition" is strategic in this context as it stresses the "becoming" of commons. "The Commons Transition Platform is a database of practical experiences and policy proposals aimed toward achieving a more humane and environmentally grounded mode of societal organization. Basing a civil society on the Commons (including the collaborative stewardship of our shared resources) would enable a more egalitarian, just, and environmentally stable society."
10. Gary Hall, *Pirate Philosophy*, MIT Press, Cambridge (Mass.), 2016, p. 20.
11. One definition of copyleft goes like this: "The right to freely use, modify, copy, and share software, works of art, etc., on the condition that these rights

be granted to all subsequent users or owners." www.dictionary.com/browse/copyleft.

12. For a background of the commons idea in Creative Commons, see Lawrence Lessig, *The Future of Ideas, the Fate of the Commons in a Connected World*, Random House, New York, 2001, written in the same period as the founding of the U.S. non-profit organization. In 2015, CC passed the 1 billion licensed works mark.

13. A 2018 documentary by Moritz Riesewieck and Hans Block: www.theverge.com/2018/1/21/16916380/sundance-2018-the-cleaners-movie-review-facebook-google-twitter.

14. Lauren Berlant, *The Commons: Infrastructures for Troubling Times*, in: Society and Space, Vol. 34(3), 2016, p. 399.

15. Ibid. p. 408.

16. Gary Hall, *Pirate Philosophy*, MIT Press, Cambridge (Mass.), 2016, p. 8.

17. Vincent Kaufmann, *Revolution in the Service of Poetry*, University of Minnesota Press, Minneapolis, 2006, p. 202.

18. If we follow Günther Anders and Elias Canetti (a thesis further theorized by Jean Baudrillard), this is because we moved beyond "point omega", symbolized by Auschwitz and Hiroshima, that made it impossible to return to normal after World War II. The common sense of chronology as a linear history was broken. The Situationist International was radical enough to embody this insight—and to act upon it. It became a "question of the generations" (Bernard Stiegler) to what extent we're still living in the shadow of this omega point. Has memory already lost its symbolic relevance and turned into empty rituals? Strong demarcation points such as 1989, 2001 and 2008 tend to overshadow the lessons of 1945. This also has implications for the avant-garde concept.

19. Quoted in McKenzie Wark, *The Beach beneath the Street*, Verso Books, New York, 2011, p. 67.

20. Both quotes from *The Beach beneath the Street*, p. 65.

21. Saskia Sassen, *Expulsions, Brutality and Complexity in the Global Economy*, Harvard University Press, Cambridge (Mass.), 2014. According to Saskia Sassen, we live in a phase "marked by expulsions—from life projects and livelihoods, from membership, from the social contract at the center of liberal democracy." (p. 29).

22. www.gregorysholette.com/.

23. www.andrewculp.org/dark-deleuze-research.

24. www.businessdictionary.com/definition/organization.html.

25. Nick Srnicek and Alex Williams, *Inventing the Future: Postcapitalism and a World Without Work*, Verso Books, London, 2015.

26. https://diem25.org/what-is-diem25/.

27. *The Beach beneath the Street*, p. 121.

28. Phrase used by Don and Alex Tapscott in *Blockchain Revolution: How the Technology behind Bitcoin is Changing Money, Business and the World*, Portfolio Penguin, London, 2016, p. 280, where they point at the contradiction of the blockchain platform Ethereum, which is "unabashedly

individualistic and private and yet depends on a large, distributed community."

29. Douglas Rushkoff, *Present Shock*, Penguin Group, New York, 2013, p. 3.

30. Peter Szondi, *Poetik und Geschichtsphilosophie*, Suhrkamp Verlag, Frankfurt am Main, 1974, p. 305, quoted in Peter Bürger, p. 92.

31. *Internationale situationiste* #12, p. 83.

Bibliography

Adilkno, *Media Archive*, Autonomedia, Brooklyn, 1998.

Adorno, Theodor W., *Minima Moralia*, Suhrkamp Verlag, Frankfurt am Main, 1951.

Aly, Götz and Roth, Karl Heinz, *Die restlose Erfassung, Volkszählen, Identifizieren, Aussondern im Nationalsozialismus*, Rotbuch Verlag, Berlin, 1984.

Barthes, Roland, *Camera Lucida*, Hill and Wang, New York, 1982.

Barthes, Roland, *A Lover's Discourse: Fragments*, Penguin, London, 1990.

Beller, Jonathan, *The Message is Murder, Substrates of Computational Capital*, Pluto Press, London, 2018.

Berardi, Franco, *Futurability, The Age of Impotence and the Horizon of Possibility*, Verso, London, 2017.

Bettelheim, Bruno, *The Informed Heart*, Avon Books, New York, 1960.

Black, Edwin, *IBM and the Holocaust*, Crown Books, New York, 2001.

Bratton, Benjamin, *The Stack*, MIT Press, Cambridge (Mass.), 2016.

Broder, Melissa, *So Sad Today*, Scribe, London, 2016.

Broder, Melissa, *The Pisces*, Hogarth Press, New York, 2018.

Broeckmann, Andreas and Knowbotic Research (ed.), *Opaque Presence: Manual of Latent Invisibilities*, Diaphanes, Zurich, 2010.

Brunton, Finn and Nissenbaum, Hellen, *Obfuscation, a User's Guide for Privacy and Protest*, MIT Press, Cambridge (Mass.), 2015.

Bürger, Peter, *Theory of the Avant-garde*, University of Minnesota Press, Minneapolis, 1984.

Calinescu, Matei, *Five Faces of Modernity*, Duke University Press, Durham, 1987.

Canetti, Elias, *Crowds and Power*, Penguin Books, Middlesex, 1973.

Capulcu Redaktionskollektiv, *Disrupt! Widerstand gegen den technologischen Angriff*, Unrast Verlag, Münster, 2017.

Chun, Wendy, *Updating to Remain the Same*, MIT Press, Cambridge (Mass.), 2017.

Cole, Teju, *Everyday is for the Thief*, Random House, New York, 2014.

Coleman, Gabriella, *Hacker, Hoaxer, Whistleblower, Spy: The Many Faces of Anonymous*, Verso Books, New York, 2014.

Culp, Andrew, *Dark Deleuze*, University of Minnesota Press, Minneapolis, 2016.

Dean, Jodi, *The Communist Horizon*, Verso Books, London, 2012.

Dombek, Kristin, *The Selfishness of Others, An Essay on the Fear of Narcissism*, Farrar, Straus & Giroux, New York, 2016.

Einstein, Mara, *Black Ops Advertising, Native Ads, Content Marketing and the Covert World of the Digital Sell*, OR Books, New York, 2016.

Fisher, Mark, *Capitalist Realism, Is there No Alternative?* Zero Books, Winchester, 2009.

Fisher, Mark, *Ghost of My Life, Writings on Depression, Hauntology and Lost Futures*, Winchester, 2014.

Foti, Alex, *General Theory of the Precariat*, Institute of Network Cultures, Amsterdam, 2017.

Freud, Sigmund, *The Interpretation of Dreams*, Barnes and Noble Books, New York, 1994.

Geertz, Clifford, *The Interpretation of Cultures*, Basic Books, 1973.

Geiselberger, Heinrich (ed.), *The Great Regression*, Polity Press, Cambridge, 2017.

Greenfield, Adam, *Radical Technologies*, Verso Books, London, 2017.

Greif, Mark, *Against Everything, On Dishonest Times*, Verso Books, New York, 2016.

Hall, Gary, *Pirate Philosophy*, MIT Press, Cambridge (Mass.), 2016.

Halpern, Orit, *Beautiful Data*, Duke University Press, Durham, 2014.

Han, Byung-Chul, *Müdigkeitsgesellschaft*, Matthes & Seitz, Berlin, 2016.

Hanru, Hou and Lonardelli, Luigia (ed.), *Please Come Back, The World As Prison*, Mousse Publishing, Rome, 2017.

Harari, Yuval Noah, *Homo Deus*, Vintage, London, 2017.

Hardt, Michael and Negri, Antonio, *Assembly*, Oxford University Press, Oxford, 2017.

Hardt, Michael and Negri, Antonio, *Commonweath*, Harvard University Press, Cambridge (Mass.), 2009.

Hartmann, Detlef, *Die Alternative: Leben als Sabotage, Zur Krise der technologischen Gewalt*, iva-Verlag, Tübingen, 1981.

Helmes, Günter and Köster, Werner (ed.), *Texte zur Medientheorie*, Reclam, Stuttgart, 2004.

Holvast, Jan, *De volkstelling van 1971: verslag van de eerste brede maatschappelijke discussie over de aantasting van privacy*, Uitgeverij Paris, Zutphen, 2013.

Houellebecq, Michel, *Whatever*, Serpent's Tail, London, 1998.

Illouz, Eva, *Why Love Hurts*, Polity Press, Cambridge, 2012.

Invisible Committee, *Now*, Semiotext(e), South Pasadena, 2017.

Johnston, Adrian and Malabou, Catherine, *Self and Emotional Life*, Columbia University, New York, 2013.

Kaufmann, Vincent, *Revolution in the Service of Poetry*, University of Minnesota Press, Minneapolis, 2006.

Keen, Andrew, *How to Fix the Future*, Atlantic Books, London, 2018.

Klibansky, Raymond, Panofsky, Erwin and Saxl, Fritz, *Saturn und Melancholie*, Suhrkamp, Frankfurt am Main, 1990.

Kobek, Jarett, *I Hate the Internet*, Serpent's Tail, London, 2016.

Langlois, Ganaele, *Meaning in the Age of Social Media*, Palgrave, New York, 2014.

Lanier, Jaron, *Ten Arguments For Deleting Your Social Media Accounts Right Now*, Vintage, London, 2018.

Larsen, Lars Bang (ed.), *Networks*, MIT Press, Cambridge (Mass.), 2014.

Lasch, Christopher, *The Culture of Narcissism*, Warner Books, New York, 1979.

Leadbeater, Charles. *We-Think*, Profile Books, London, 2008.

Lepenies, Wolf, *Melancholie und Gesellschaft*, Suhrkamp, Frankfurt am Main, 1998.

Lévi-Strauss, Claude, *Der Weg der Masken*, Suhrkamp, Frankfurt am Main, 2004.

Lévi-Strauss, Claude, *Tristes Tropique*, Penguin Classics, London, 2011.

Löffler, Petra, *Verteilte Aufmerksamkeit, Eine Mediengeschichte der Zerstreuung*, Diaphanes, Zürich, 2014.

Lottman, Herbert R., *Albert Camus, a Biography*, Gingko Press, Corte Madera, 1997.

Lovink, Geert, *Dark Fiber*, MIT Press, Cambridge (Mass.), 2002.

Lovink, Geert, *My First Recession*, NAi, Rotterdam, 2003.

Lovink, Geert, *Zero Comments*, Routledge, New York, 2008.

Lovink, Geert, & Schultz, Pit, *Jugendjahre der Netzkritik*, Institute of Network Cultures, Amsterdam, 2010.

Lovink, Geert, *Networks without a Cause*, Polity Press, Cambridge, 2011.

Lovink, Geert, *Social Media Abyss*, Polity Press, Cambridge, 2016.

Lovink, Geert, and Rossiter, Ned, *Organization after Social Media*, Minor Compositions, Colchester, 2018.

Mause, Lloyd de, *Reagan's Amerika*, Stroemfeld/Roter Stern, Frankfurt am Main, 1984.

McAfee, Andrew and Brynjolfsson, Erik, *Machine Platform Crowd*, Norton & Company, New York, 2017.

Mejias, Ulises, *Off the Network, Disrupting the Digital World*, University of Minnesota Press, Minneapolis, 2013.

Monbiot, George, *Out of the Wreckage*, Verso Books, London, 2017.

Motyl, Katharina and Schober, Regina (ed.), *The Failed Individual*, Campus Verlag, Frankfurt, 2017.

Nagle, Angela, *Kill All Normies*, Zero Books, Winchester, 2017.

O'Neil, Cathy, *Weapons of Math Destruction*, Penguin, London, 2016.

Nevejan, Caroline, *Presence and the Design of Trust*, University of Amsterdam PhD, 2007.

Pfaller, Robert, *On the Pleasure Principle in Culture: Illusions Without Owners*, Verso Books, London, 2014.

Preciado, Paul B., *Testojunkie*, Feminist Press, New York, 2013.

Rid, Thomas, *Rise of the Machines, the Lost History of Cybernetics*, Scribe, Brunswick, 2016.

Rossiter, Ned, *Software, Infrastructure, Labor, A Media Theory of Logistical Nightmares*, Routledge, New York, 2016.

Sassen, Saskia, *Expulsions, Brutality and Complexity in the Global Economy*, Harvard University Press, Cambridge (Mass.), 2014.

Schnitzler, Hans, *Kleine filosofie van de digitale onthouding*, De Bezige Bij, Amsterdam, 2017.

Sontag, Susan, *Under the Sign of Saturn*, Vintage Books, New York, 1981.

Srnicek, Nick and Williams, Alex, *Inventing the Future: Postcapitalism and a World Without Work*, Verso Books, London, 2015.

Srnicek, Nick, *Platform Capitalism*, Polity Press, Cambridge, 2016.

Stiegler, Bernard, *Technics and Time 2: Disorientation*, Stanford University Press, Stanford, 2009.

Stiegler, Bernard, *Uncontrollable Societies of Disaffected Individuals*, Polity Press, Cambridge, 2013.

Stiegler, Bernard, *The Lost Spirit of Capitalism*, Polity Press, Cambridge, 2014.

Stiegler, Bernard, *Automatic Society, The Future of Work*, Polity Press, Cambridge, 2016.

Stiegler, Bernard, *The Neganthropocene*, Open Humanities Press, London, 2018.

Stocker, Gerfried and Schöpf, Christine, *Ars Electronica 96: Memesis - The Future of Evolution*, Springer, Wien, 1996.

Tapscott, Don and Alex, in *Blockchain Revolution: How the Technology behind Bitcoin is Changing Money, Business and the World*, Portfolio Penguin, London, 2016.

Tufekci, Zeynep, *Twitter and Tear Gas: The Power and Fragility of Networked Protest*, Yale University Press, London, 2017.

Turkle, Sherry, *Reclaiming Conversation: The Power of Talk in a Digital Age*, Penguin Press, London, 2015.

Vaidhyanathan, Siva, *Anti-Social Media*, Oxford University Press, New York, 2018.

Wallman, James, *Stuffocation: Living More With Less*, Penguin Books, London, 2015.

Wark, McKenzie, *The Beach beneath the Street*, Verso Books, New York, 2011.

Wilder, Darcy, *Literally Show Me a Healthy Person*, Tyrant Books, New York, 2017.

Žižek, Slavoj, *The Year of Living Dangerously*, London, Verso Books, 2012.

Žižek, Slavoj, *The Courage of Hopelessness*, Allen Lane, London, 2017.